Quarrying and Rockbreaking

The Operation and Maintenance of Mobile Processing Plants

by D. Lester M.B.I.M., A.I.Q.

An Intermediate Technology Publication

© Intermediate Technology Publications, 1981

Published by Intermediate Technology Publications Ltd,
9 King Street, London WC2E 8HN, U.K.

ISBN 0 903031 80 9

Reproduced from copy supplied
Printed and bound in Great Britain
by Billing and Sons Limited
Guildford, London, Oxford, Worcester

ACKNOWLEDGEMENTS

The printing of this publication has been made possible by grants from the following: the Priest and People of Our Lady and St. Oswin's R.C. Parish, Tynemouth; the Episcopal Church in Scotland (Action for World Development Committee); Dartington Hall School; St. Michael's Parish Church, St. Albans and an anonymous donor. The Intermediate Technology Development Group gratefully acknowledges their generosity.

The author gratefully acknowledges the assistance given by Pegson Ltd., Coalville, U.K., in giving access to their technical records and library; and by Demister A.B., Malmo, Sweden, in the preparation of the material on pages 88–95. The material on pages 95–104 contains extracts from service manuals reproduced by courtesy of the Universal Conveyor Co. Ltd. Leicester U.K.

A MOBILE ROCKCRUSHING PLANT AND A WHEELED LOADER IN OPERATION

CONTENTS

1

PREFACE

Broken rock in the form now known as "aggregates" has been needed for many years for a variety of community requirements. Initially rock was needed for repairing and draining tracks and pathways which were later upgraded into firm, level roadways by surfacing with broken, sized rock.

As Portland or "hydraulic" cement became readily available demand for aggregates for the production of concrete developed. Concrete began to be used for such purposes as house floors, foundations and walls and in precast form for building blocks, roof tiles, paving slabs and drainage pipes. For satisfactory concrete mixes the rock had to be broken into small, accurately graded sizes. Uncrushed natural gravel could often be used but where there were no deposits of suitable gravel, solid rock in boulder form or from a cliff face had to be extracted and processed into usable particle sizes.

In practical terms, supplies could only be won from such sources where the rock was weathered, naturally fissured or comparatively soft.

The rock had first to be extracted in lump form of a size and weight which could be carried by one person or by two using a sledge or cradle. For this first stage, heavy hammers and steel bars and wedges were used. These latter could often be made from old motor half-shafts forged to a chisel shape for the bars or tapered for wedges. Half-shafts could also be forged into "jump" drills; with these a line of holes could be hammered into solid rock and wooden plugs inserted. The plugs were then soaked with water, which would make them expand and split the rock.

The lumps were then carried to a central area with a firm, flat surface and broken down into usable, graded sizes.

Lighter tools — "whip hammers" — were used and as the lumps were progressively broken down the particles were each passed through metal rings, perhaps old steel washers, to size them and were eventually put aside in piles of the usable sizes needed, perhaps $\frac{1}{2}''$ (12 mm), 1″ (25 mm) and $1\frac{1}{2}''$ (38 mm).

The broken, graded rock would be sold to a contractor or used for community purposes.

3

These methods are still used in some areas and can provide employment and income for entire families or family groups.

Output is, however, very low and the limited quantities of aggregate which can be produced are not nearly sufficient to meet the needs of an expanding, developing community requiring roads, buildings, housing and drainage. There may also be a requirement for industrial buildings and for civil engineering works such as bridges and water-works requiring thousands of tons of good quality aggregates. Farming and agricultural activities generally will also generate a demand for concreting aggregates.

Mechanical production thus becomes necessary. In the early stages of mechanisation, small stone crushers with mouth openings up to, say, $20'' \times 10''$ ($510\,mm \times 250\,mm$), driven by petrol or diesel engines of around 20 b.h.p., can be used. They are generally hand-fed and very simple but their outputs of approximately 15 tonnes per hour are still insufficient to meet the present needs of all but the smallest communities. The modern generation of rock processing plants can produce up to and in excess of 300 tonnes per hour, depending on the size and type of plant chosen. Inevitably the satisfactory operation of such machines and of the ancillary activities involved calls for technical capabilities which may not, in the early stages at any rate, be locally available.

The book is therefore an attempt to bring together some of the many considerations and requirements of a fairly high output rock processing undertaking and should help owners of such plant and their operators to gain the best returns from the capital and labour invested.

There is no attempt here to evaluate the capital needed or the sources available. Competent financial advice and adequate investment support should be sought in the early stages of the undertaking and it is worth remembering that some of the equipment which will be needed, for example bulldozers and rock drills, can often be obtained on contract hire, with or without operators.

Finally the reader should bear in mind that this book supplements and does not replace the service manuals which he will receive from the suppliers of his equipment. Those manuals must always be consulted for detailed information on the operation and maintenance of any particular machine.

ENTER DETAILS OF YOUR PLANT:

Primary unit

Plant No._____

Feeder No. _____

Crusher No. _____

Secondary unit

Plant No._____

Screen No._____

Crusher No._____

Tertiary unit

Plant No. _____

Screen No. _____

Crusher No. _____

Additional
screening unit

Plant No._____

Screen No. _____

Portable
*conveyors*_____

_____

_____

Power unit

Engine No._____

Alternator No. _____

When ordering spare parts always quote this information.

INTRODUCTION

This manual is intended as a guide to the day-by-day operation and maintenance of a rock crushing and screening plant and it starts with a brief consideration of quarry face activities and internal transport.

There is no exact science of quarrying and this manual is not a text-book. Bear in mind that, particularly at the rock face, every operation is likely to dictate its own empirical solution. The ideal approach to problems of winning and processing the rock will change with the life of the rock face and of the plant; knowledge and experience will accumulate and the conscientious manager will take care to record the developing operational demands of his quarry.

In establishing a new quarrying operation there will of course be strictly commercial considerations which are not the direct concern of this book but which will have to take account of the detailed answers to such questions as:

(1) What material is needed by the local market? Is it dry stone? Coated stone? Sand?
(2) To what extent, having in mind the material available, can you as the quarry operator influence the demand?
(3) What material is available?
(4) Where does the demand exist in relation to the available source or sources of material?
(5) For how long will there be market demand and how does it compare with the quantity of material available?
(6) What machines are best suited to processing and handling the material?
(7) How efficient is the local parts and service back-up for the machines selected?
(8) If you must employ inexperienced labour what facilities are available locally for training, particularly by the suppliers of your selected machines?
(9) Having in mind the nature of the material, what will be the cost of producing aggregates which the accessible market will accept?

The answers to all of these questions will be influenced by your knowledge of aggregates, their sizes, shapes, workability, textures etc., and the next section is a general consideration of these factors.

Aggregates may be simply defined as any broken hard material, not necessarily of natural origin, or fragments and particles of such material.

They may be used with admixtures, for example in concrete or asphaltic concrete or "dry" as in road bases and for drainage.

The eventual employment of the aggregates will certainly dictate some consideration of their physical characteristics and the market may well impose its requirements on the processing plant to be employed in producing saleable aggregates from the available raw material.

Quality requirements of aggregates

Quality requirements are likely to involve:

(1) *Grading and sizes*

The strength of structural concrete, of asphaltic concrete for road paving and of dry stone bases depends largely on the sizes and distribution i.e. grading of the aggregate particles. Workability and density under compaction will be directly determined by sizes and grading and consistent control of these factors at the plant is vital.

Sieve analysis tests may be demanded for the evaluation of sizes and grading.

(2) *Particle shape*

Workability, interlocking and density under vibration or compaction will be influenced by the shape of the particles.

Particles with rounded faces will not interlock and although easily workable will give low density due to large voids.

Flaky particles will interlock satisfactorily with good density and reasonable workability but are liable to crack under compaction which will spoil the aggregate grading.

Cubical shapes should be aimed at and will produce the best results in all these requirements. Again testing may be called for and requires the use of a calliper to measure the longest and smallest dimension of each particle. The ratio of longest to smallest dimension ought not to exceed 3 for more than 5% of the sample or 2.5 for more than 20% of the sample.

(3) *Cleanliness*

So that the aggregate particles can be adequately coated with asphalt, bitumen, tar or cement they must be clean, otherwise the bond will be lost. Dust and clay, particularly when adhering to the aggregate, are the main contaminants and clay will also affect the self-draining characteristics of dry stone bases.

Sampling and testing to specified requirements will probably be demanded.

(4) *Surface texture*

This is important where the aggregate is used for asphaltic pavements and will affect the non-slip characteristics of the traffic surface, the initial absorption of asphalt and the subsequent absorption of water which may eventually lead to failure of the road surface.

Testing of surface texture is usually by visual inspection.

(5) *Absorption*

Absorption is generally related to surface texture and is important in the determination of the quantity of asphalt needed to achieve specified mixes.

It can be tested by the wet and dry weighing of samples.

(6) *Toughness*

This is important in the determination of the strength of the concrete or asphaltic mix, resistance to damage or deformation during construction and the resistance of a road surface to wear by traffic.

The Los Angeles abrasion test is often specified.

(7) *Soundness*

This is a measure of the resistance of the aggregate to decomposition under the action of natural conditions such as rain, wind, frost and soundness tests may be specified.

(8) *Affinity for asphalt*

Some aggregates, typically quartzites, have unsatisfactory coating characteristics resulting from highly polished fracture faces. Such particles will strip off a compacted asphaltic surface and the pavement will break up.

Again the client may specify tests.

The origins of aggregate rocks

Your appreciation of the qualities and suitability of aggregates will be helped by knowledge of their origins. These are three main classifications of rock:

8

(1) *Igneous rocks*. These were formed by the cooling and crystallisation of molten materials thrust from the core of the earth into its mantle and they comprise a high proportion of the earth's crust, visible or concealed. There are two further classifications:

(a) Extrusive. These rocks cooled rapidly in the atmosphere and the process still occurs in volcanos. They are close grained and hard but sometimes brittle. Examples are Basalt, Rhyolite and Andesite.

(b) Intrusive. Such rocks cooled more slowly beneath the surface, are of coarser grain and generally hard. The coarse grain size can lead to brittleness. Common types are Granite, Gabbro and Diorite.

(2) *Sedimentary rocks*. These were initially deposited as a variety of fine grains often resulting from the erosion and weathering of exposed surface rocks. Volcanic ash was another source. Grains also resulted from the decomposition of plants and animals, especially vertebrates and shell-forming creatures. The succeeding layers of grains were cemented and compressed eventually forming sedimentary rocks in which fossils are fairly frequent. Examples are Sandstone, Limestone and DoComite. Old sandstones can be of adequate strength for aggregates but may be crushed by rolling during road construction and may need testing for affinity for asphalt.

Limestones are often hard although not as hard as igneous rocks and the darker colours are usually the hardest of the classification. Affinity for asphalt is good.

Quartzites are often very hard but may be brittle and as indicated earlier have poor coating qualities.

(3) *Metamorphic rocks*. Sedimentary and igneous rocks were changed in structure by pressure, often combined with heat during disruption of the earth's mantle and became metamorphosed. Sandstones were changed into forms of Quartzite, Limestones became Marble and Shales became Slates. Gneisses and Schists were among the metamorphic forms originating in igneous rocks. Metamorphosis usually made the original rocks harder but more brittle.

The sources of aggregate rocks
Workable locations are likely to be found in:

9

(1) Rock outcrops, which may provide either bedrock or boulders often covered by a varying depth of overburden sand, laterite or soil. Exposed rock surfaces may be weathered or to some extent decomposed. Bedrock will usually be fissured or laminated.

(2) Sand and/or gravel pits which will probably be contaminated by clay and silt requiring removal by washing the aggregate. Sands may need processing to separate the varying grain sizes. Gravel aggregates won from pits will often be angular in shape with rough surfaces.

(3) Beaches can also yield gravels and sands and here the particles will be rounded in shape. Sands will often be single sized and all aggregates will require washing to eliminate salt.

(4) River beds, either dry or submerged, will yield sands, gravels and often boulders. There may be clay and silts. The particle surfaces will generally be rounded and smooth. Sands will be encountered in varying grain sizes and may need separating.

Processing rock into acceptable aggregates
Most naturally occuring rock sources are affected by one or more defects which must be remedied in the processing plant.
These defects are commonly:

(a) The presence of unwanted impurities as frequently found in sand and gravel deposits. These impurities may be, for example, vegetation fragments, clay, silt, salt and are usually removed by washing.

(b) Unacceptable shape or surface texture, usually corrected by one or more crushing stages.

(c) Oversize particles, which can be separated by screening and if the proportion is high enough for economic handling can be crushed and combined with the remaining product.

(d) Undersize particles which again can be separated by screening or, in the case of silts, by washing combined with the screening.

As discussed earlier, particle shape is important and to achieve a cubical shape with angular faces it will almost always be necessary to crush the aggregate, particularly when it is excavated from gravel deposits.

Locating and evaluating a workable quarry site

The factors to be considered in relation to a semi-permanent (i.e. mobile or transportable) plant may be somewhat less limiting as compared with a permanent installation with a long projected operational life.

Under this heading we are not concerned with the more or less temporary employment of a mobile plant to boost the output of an established quarry.

Having identified and evaluated the demand for aggregates, either for the private contractor or for a State project, a search will have to be made for a convenient and accessible deposit of suitable rock. If the likely area has already been surveyed a contour map can be an initial reference. The map will show promising hills and close contour lines will identify such useful features as steep slopes, spurs and cliff faces. Sometimes the map may incorporate a geological survey which will help in identifying suitable outcrops.

A physical search can follow, remembering that in many countries likely sites may be under jungle, committed grazing land or estate cultivation. Local knowledge, particularly in logging concessions, can be invaluable. Each potential site will have to be evaluated in terms of:

(a) The quantity of rock which can be made available to the processing plant. For a mobile plant smaller reserves than for a permanent installation can usually be considered although there may well be an eventual intention to develop the operation into a larger scale undertaking.

(b) The quality of the material in terms already discussed.

(c) The relation of the site to such ancillary facilities as fuel supplies, engineering workshops, potable water, processing water etc.

(d) The length and slopes of the access road.

(e) The availability of electric power. This will determine the method of driving the processing plant and will indicate the necessity or otherwise of installing a generator set.

(f) The availability of flat or gradable land, not only for the processing plant but also for workshops, mess buildings, office, secure store, fuel storage and most importantly stockpiles with convenient vehicle and loading access. This latter requirement can hardly be over-emphasised.

11

(g) Overburden disposal, remembering that an eventual requirement may be reclamation of the site by spreading the overburden and top soil in correct placing.

(h) The local impact of the quarry on agricultural land, inhabited areas, tourist and public access generally. Safety and security from intrusion will have to be taken into account.

(i) Land title and Royalty calls.

(j) Local availability of labour, particularly of competent mechanics.

Indigenous cultures and traditions may also compel other considerations.

Territorial regulations

You will have to identify and study any State regulations and requirements relating directly or indirectly to quarrying activities, mineral extraction, machinery, safety, health and welfare. There will certainly be regulations concerning blasting operations which will be the responsibility of the management. There may be a requirement to notify one or more Inspectorates and perhaps Police/Security Authorities. The local Administrative Authority will also be interested in your activities.

Nothing in this manual is to be construed as in any way qualifying the regulations and statutory requirements affecting your particular operation.

MECHANICAL ROCK CRUSHING ON A SMALL SCALE

A comparatively small, simple plant capable of producing around 20 tonnes per hour of crushed rock in sizes $\frac{1}{4}''$ (6 mm) to $\frac{3}{4}''$ (19 mm) will often meet local demand for block making etc.

Two-stage crushing is preferable to single stage and the mechanical plant might comprise:

(a) A single toggle jaw crusher, size 380 mm × 610 mm, as the primary machine. This will require approximately 35 b.h.p. (26 kW) to drive.

(b) A single toggle jaw granulator, size 150 mm × 910 mm, as the secondary machine, requiring approximately 50 b.h.p. (38 kW).

Preferable to a jaw granulator but at greater initial cost would be a cone crusher, size 610 mm, requiring approximately 30 b.h.p. (22 kW).

(c) A belt conveyor between the primary and secondary crushers. The length will vary with the space available and with the level of the terrain and the drive will probably need approximately 5 b.h.p. (4 kW).

(d) A second, similar belt conveyor between the secondary crusher and the screen.

(e) A sizing screen which could either be a mechanical vibrating screen with 3 or perhaps 4 decks, nominal size 2.5 m × 1.3 m, or a cylindrical screen with perhaps 4 sections and overall size 1 m diameter × 5 m long. In either case allow approximately 7.5 b.h.p. (5.5 kW).

(f) A third belt conveyor, similar to C and D to return oversize stone from the screen to the secondary crusher for recrushing.

Each machine can have its own, separate power unit, preferably an electric motor, and the drives can generally be by flat leather belts although a vibrating screen will, ideally, need vee belts.

Alternatively a single power unit, a diesel engine or an electric motor might be employed, in which case the individual machines will be driven through an arrangement of pulleys and countershafts.

Suitable second-hand equipment can often be found locally and adapted to drive the machines. Motor car gearboxes make good speed reducers for the conveyors and a complete car engine with its transmission will drive a vibrating or cylindrical screen although it will have to be protected from dust.

Foundations for the crushers will have to be made of concrete and they should be sited to make use, if possible, of any ground slope to provide height for the final storage bins or compartments.

Other structures, particularly for the conveyors, can be made of timber, often in the round. Timber can also be used to construct small storage bins for the overhead loading of lorries or for sloping compartments from which crushed stone can be shovelled into lorries.

The floor and sides of the primary crusher dump hopper or chute will have to be made of concrete or steel and must be as strong as you can make them. At the bottom there should be an adjustable door and a safe platform from which a man can control the flow of rock to the primary crusher. Give this man room to do his job safely and in particular ensure that he is protected from (a) stones entering the chute from the lorries and (b) the revolving flywheels of the crusher.

If you decide to use mechanical means of feeding rock into the primary crusher a plate apron feeder can, with ingenuity, be constructed from old crawler tracks, sprockets and rollers.

When planning and building a small plant such as this, try to make sure that the path of the stone through the processes is, as far as possible, a straight line, both vertically and horizontally. Any change of direction increases both the rate of wear and the consumed horsepower. For example if a belt conveyor elevates a flow of stone 5 metres and then drops the stone 2 metres into the secondary crusher (a) some of the horsepower consumed by the conveyor is being wasted and (b) there will be unnecessarily heavy wear on whatever part of the crusher the stone hits at the bottom of the drop.

Remember too that stone does not drop vertically from the head drum of a belt conveyor. It is thrown forward and downward in a curve or trajectory which depends on the speed of the belt, the angle of the conveyor and the size of the stone. You must allow for this trajectory in locating the machines relative to each other and it is often possible to find the best position for the head drum by trial and error.

It is however possible to plot the curve of the trajectory so that the position and dimensions of the chute or receiving hopper can be planned at an early stage in building the plant.

Wherever the flow of stone collides with the side or bottom of a chute, try to incorporate a "dead pocket". This is a ledge which can be made of strong timber and on which stone remains to protect the chute from wear.

The rock to feed a plant such as this will have to be pre-broken to a lump size of approximately 12″ (300 mm). In the early stages there may well be loose boulders on the site; later it will almost certainly be necessary to blast from bedrock. In either case the pre-breaking to 12″ lump size will probably be done by hand hammer by groups working on the quarry floor and the number of people so employed will determine (a) the rate at which the plant can be fed and (b) the total labour requirements for the project. If the crushing plant is located at approximately the safe minimum distance from the quarry floor i.e. 250 metres, one lorry will be sufficient to keep the plant fed at a regular rate with lump rock.

Assuming that drilling and blasting are done by a contractor, as is often the case, the likely total direct labour requirements will be:

Prebreaking rock on the quarry floor, 10 men.

Hand loading the lorry, 4 men.

(usually these two groups will help each other)

Driving the lorry, 1 man.

Controlling the feed to the primary crusher, 1 man.

Cleaning up spillage and generally tending the plant, 1 man.

Tending the storage bins and supervising the loading of vehicles collecting crushed rock, 1 man.

(If storage compartments at or near ground level are used, this will probably need 2 men)

Running maintenance of the lorry and the crushing plant, welding, greasing etc. 1 man.

(If the plant is electrically driven and this man is not also an experienced electrician, then the services of a second maintenance man, suitably experienced, will be needed.)

Site clerical work, recording, invoicing, ordering etc., 1 man.

Site manager/supervisor, 1 man.

PART ONE: QUARRYING AND PROCESSING

THE QUARRY AREA AND ROCK FACE

The majority of operations are associated with extraction from bed-rock sources and we are assuming that a new quarry face will have to be developed. Probably the first large plant item on site will be a bulldozer which will:

(a) Cut and grade the access road to the site.

(b) Cut an access road to the area which will become the head, or top, of the quarry face.

(c) Carry out the initial levelling and grading of the area assigned to the plant and stockpiles.

(d) Remove at least sufficient overburden to allow an early start on developing the largest practicable rock face. The overburden should not now or subsequently be dozed over the quarry face onto the quarry floor *except that* in the very early stages it may be useful in levelling the plant area. Even at slightly higher cost the overburden should be piled or perhaps spread at a location where it will not subsequently have to be rehandled except for reclamation purposes.

(e) During the operations, particularly (c) and (d) the bulldozer should if possible prepare a ground stock of boulders and loose rock, probably weathered and substandard, which can usefully form the first feed material to the crushers and screens. This rock will test the plant arrangement, giving an opportunity to make initial adjustments, will provide initial experience for the operators and when processed can be used on the access and internal roads. In these early stages the bulldozer will probably be hired, with or without a driver/mechanic.

Security
At this stage the site is becoming potentially hazardous and most operators will consider security. The area should be securely fenced, warning notices erected and public access roads gated.

Explosives

Early consideration will have to be given to the safe storage of explosives and here it is vitally important to identify and comply with all State and local regulations, certification etc.

The following points are for guidance only and must be checked against the regulations enforced in your area, bearing in mind that the prime requirements are (a) to protect the explosives from deterioration (b) to eliminate access by unauthorized persons (c) to reduce the risk and potential effects of accident.

(1) Explosives must never be stored with other commodities.

(2) The location of the magazine (explosive store) will be determined by local regulations but should not be less than, say, 20 metres from any public facility or access.

(3) Detonators and fuses must be stored in a separate magazine or in a secure annex to the main magazine.

(4) The magazine must be strongly built of permanent materials e.g. brick or concrete and must have a lockable steel sheathed door opening outwards. It must not be possible for hinges, fastenings etc. to be dismantled from the outside.

(5) The magazine, including the internal face of the door, must be wood lined. A sun roof should be incorporated in hot climates. The floor should be asphalted.

(6) Storage tiers must be so arranged that identification marks on cases can be seen, particularly the manufacturing dates. This will facilitate the issue of explosive in date sequence.

(7) There should be a secure boundary fence around the magazine area with one lockable gate and warning notices.

(8) An encompassing earth rampart at least as high as the building should be considered. In this case there may be a drainage requirement.

(9) There must be a lightning conductor.

(10) All internal fittings, including tools, must be of copper, brass or wood.

(11) Where ventilation holes are required the pairs on inner and outer cavity walls should be offset and not directly aligned. The outside holes should be louvred and the inside holes fully covered by brass or copper gauze of approximately 60 mesh.

(12) Timber should be suitably treated to inhibit fungus or insect attack.

(13) The local security authority may demand the employment of a night watchman, to be in attendance also in daylight hours when the site is otherwise unoccupied.

Procedural rules for explosives

Access to the magazine should be limited to nominated personnel and there should be written and posted rules to cover such requirements as:

(1) The handling of explosives which must be carried out with great care and in an unhurried manner.

(2) No materials other than explosives must be stored in the magazine.

(3) Opening bulk cases and removing explosives must not be done inside the magazine. A detached, separate building should be provided for this purpose.

(4) No matches, cigarette lighters or any naked flame should be taken in or near the magazine or unpacking station.

(5) All access keys must be kept in a safe in a separate, locked and supervised building and issued only against the signature of a nominated person.

(6) Rubber or leather overshoes, sewn not nailed, should be provided in the magazine and used in such a manner that (a) outdoor shoes are never in contact with the magazine floor; (b) the overshoes are never in contact with the outside surface. This is to exclude grit from the magazine.

(7) Stocks must be issued and used in date sequence.

(8) There must be a system of recording incoming stocks and issues.

(9) Good "housekeeping" around the magazine is essential. There must be no nearby accumulation of scrap metal, timber, old tyres or any inflammable material including vegetation.

(10) *All* packing materials, boxes, wrappers etc. should be taken to a remote, designated area and burnt after a thorough check to ensure that no explosives have been overlooked.

(11) Explosives and detonators or fuses must never be carried in the same container. Detonators must never be carried in the pocket. A proper detonator case or, if they are still

in their original packages, a strong canvas or leather bag, must be provided.

(12) As far as possible explosives must be protected from direct sunlight and rain or snow. Shade or cover must be provided as necessary.

(13) Small quantities of explosives may be conveyed to the charging area on the quarry face in canvas or leather bags restricted to that purpose. Larger quantities should be taken to the face in the original complete cases and the drill holes loaded directly from the cases. Wrapping materials and empty cases must not be allowed to accumulate in the charging area and must be disposed of as in (10).

(14) A responsible foreman or manager should at frequent regular intervals inspect, approve and countersign all aspects of the storage, issue and application of explosives.

No matter how comprehensive the storage conditions, explosives are liable to deterioration and may have to be disposed of safely other than by shot-firing on the quarry face. When this happens it will almost certainly be obligatory to notify the security authority or police and the advice and assistance of the manufacturers or agent should be requested before any action whatever is taken.

In all matters relating to explosives be conscious of your statutory obligations and be sure that all personnel are similarly aware.

Developing the rock face
Every site will dictate its own solutions to problems of working methods and in the early stages at any rate these solutions will probably be far from ideal. Compromises will have to be accepted.

Economic production will call for "benches" on the rock face. Initially these benches will provide safe platforms for drilling. Eventually they will also be used by loading and hauling vehicles. The aim will be to provide reasonably smooth, level and wide working areas and the quarry floor, as it advances below the benches, must also be kept level, reasonably smooth and free of loose rock except as ground stocks. It will usually be necessary to create a sump for drainage and to provide pumps and a pipe line for disposal of ground water.

Benches of limited height will probably be the initial aim, eventually to be combined to a convenient working height and width as mainly dictated by the total height of the face. In this process the first blasts will almost certainly yield weathered degraded rock, useful after crushing for access and internal roads.

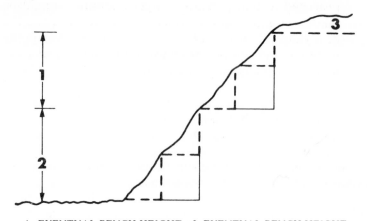

1. EVENTUAL BENCH HEIGHT   2. EVENTUAL BENCH HEIGHT
3. OVERBURDEN
Fig. 1   DEVELOPMENT OF TWO BENCHES

Drilling

The early drilling will probably be possible with jackhammer drills since the holes will be of short length and small diameter. Selection of the production drilling equipment will depend on a variety of local factors and expert opinion should be sought before making a final decision.

There is an extensive choice and a knowledge of descriptive terms is useful:

(a) Benchers are drilling machines for small diameter long holes. The supporting and feeding column is secured to the rock surface by anchor bolts.

(b) Jackhammers are hand held for small diameter holes, long or short, often used for secondary drilling of slabs and boulders following primary blasts.

(c) Wagon drills are for long holes of large diameter. They may be trolley mounted or self propelling on crawler tracks driven by air motors and capable of towing a compressor.

20

(d) Rotary drills press and rotate the drill bit against the bed-rock. Flushing of the hole i.e. the extraction of rock fragments and dust is by water or air. The method may be used either with a hollow drill bit set with a ring of industrial diamonds to extract a core of rock or with a full face bit for normal production holes.

(e) Rotary percussive drills combine hammer blows with rotation.

(f) Hammer drills employ only percussive blows to fragment the rock with interrupted rotation, used only for the constant re-alignment of drill faces in contact with the rock.

(g) Down-the-hole-drilling. This is similar to hammer drilling but instead of the hammer blows being applied to the end of the drill rod at the surface the hammer is at the bottom of the hole with the drill bit and follows it down.

(h) Integral drill steels comprise one rod of fixed length. The operation starts with a steel of, say 100 cm length, replaced in sequence by progressively longer steels up to a maximum of approximately 700 cm. They usually have chisel tips.

(i) Extension drill steels can be used for holes deeper than those possible by integral drill steels. The rods are available in varying lengths up to 7 metres, although most quarries use a maximum length of 3 metres, and are successively coupled together with the drill bit remaining coupled to the lower end.

(j) Drill bits are available in a variety of forms, the most common for quarrying being:

  (i) Cross bits on which four tungsten carbine tips are brazed at 90° alignments around a central flushing hole with two or four additional flushing holes between the tips. The usual diameters for quarrying are from 36 mm to 50 mm.

  (ii) X-bits differ slightly in that the tips are not at right angles to each other. They have an advantage over the cross bit which due to the grit blown by the flushing tends to wear to a "square" shape across the diameter. The asymetrical arrangement of the four outer flushing holes in the X-bit almost eliminates this progressive wear pattern but the tips are more difficult to grind when worn.

The maintenance of drill steels and bits is important, particularly from the point of view of cost and the workshop should include a special-purpose grinder. Usually one man will be employed full time on grinding and if he is not initially skilled the manufacturers or agent will generally co-operate in training him.

Selection of the production drilling equipment

Factors for early consideration are:

(1) A quarry face feeding a semi-permanent processing plant is likely to be of limited production life and comparatively small in extent. It will probably not be practicable to develop benches high enough and wide enough to justify the capital cost of a wagon drill.

(2) Water is more efficient than air for flushing small diameter holes, particularly if they are more than 3 metres long. Water will also reduce wear on the bits. Therefore a supply of water is desirable.

(3) The nature of the rock face or outcrop may dictate the use of a jackhammer drill i.e. on rough terrain or when handling boulder and slab outcrops it may not be possible to manoeuvre the larger drilling gear economically.

(4) The size of the primary crusher will determine the extent to which the rock is to be fragmented and the amount of secondary breaking on the quarry floor. This secondary breaking might have to be by secondary drilling and blasting, plaster shooting or drop-balling and a saving in the cost of primary drilling and blasting might be offset by the subsequent breaking.

(5) Small diameter holes are drilled at closer intervals than larger diameter holes. This closer space results in more even fragmentation and a smaller lump size on blasting.

(6) If the bedrock is fissured or heavily weathered long hole drilling may well be difficult and the blast results unsatisfactory. The holes may be difficult to load leading to misfires.

Consideration of these factors in relation to the comparatively small quarrying operation with which this manual is concerned usually leads to the selection of benchers for the main drilling on the face with jackhammers used to supplement the benchers and for secondary drilling of large boulders or slabs.

Both types of drill are powered by compressed air and one or more air compressors, probably truck or trolley mounted, will be needed. For maximum drilling efficiency, air at around 85 to 100 lbs per square inch will be needed and each drill will take approximately 60 c.f.m. (cubic feet per minute). The length of the air hose must be kept as short as practicable for better maintenance of the pressure and to minimise accidental damage. It will nevertheless probably be convenient to operate two drills from a compressor of, say, 130 c.f.m. The air from the compressor can be used for flushing the holes as they are drilled but remember that water flushing is to be preferred. A storage tank supplied by a rising main through which the water will probably have to be pumped is well worth consideration.

Primary drilling
Whatever the early nature of the terrain, the drilling crew, under the direction of the management, should work towards establishing a pattern of drilling which will help the quarry site to operate safely and economically. This pattern is different for every rock face and will, in fact, change as a particular face develops. The aim should always be the eventual development of the highest, longest face which the quarry contours allow and wherever possible random drilling in the interests of immediate though limited crushing production should be resisted.

Records should be maintained to show the yield of rock per kilo of explosive and in a small quarry the optimum will probably be around 4 to 6 tonnes.

The drill size and hence the hole diameter will of course have to match the available explosive cartridge size and for the smaller operation 32 mm will probably be a convenient diameter. Depending on the explosive chosen 32 mm diameter holes will charge at approximately 1.3 kg per linear metre.

Most drilling charge-hands have their own formulae for determining the optimum drilling pattern for a given rock face. A typical formula states that:

$$C = \frac{S \times B \times D}{6.4}$$

Where C is the charge in kgs of explosive per hole.

S is the spacing in metres
B is the burden in metres
D is the depth in metres.

These factors as applied to bench drilling are:

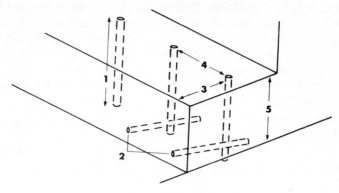

1. 6 METRES   2. TOE HOLES   3. BURDEN 1.5 M   4. SPACING 1.5 M
5. BENCH HEIGHT
Fig. 2

For example, if spacing and burden are set at 1.5 m each, then for a hole depth of, say, 6 m, the explosive charge per hole will be:

$$C = \frac{1.5 \times 1.5 \times 6}{6.4} = 2.1 \text{ kgs.}$$

In a 32 mm diameter hole, this charge will occupy approximately 1.6 m. The top of the hole must be closed by stemming which can be, for example, sand, clay or quarry dust, in all cases free of angular fragments which might cut the fuse when tamped and always used damp. The stemming depth should be around twice the burden − in this case 3 m. Hence there is a hole length of 3 m for a charge occupying 1.6 m and for the best fragmentation result, the charge will be loaded in two parts separated by more stemming.

The holes will generally, in the early stages of development, be drilled vertically and to a depth of from 0.5 m to 2 m below the quarry floor.

This "extra" depth will be determined by local conditions and will vary with the amount and direction of fissuring on the face. As the benches are formed, slanted holes will provide a safer, more stable face although fissuring of the bed rock might lead to difficulty in maintaining the inclination of the holes. The angle will usually be between 5° and 15° and in addition to preventing overhang, inclined holes will generally help breakout at the toe of the face.

Breakout will also be improved by horizontal toe holes drilled close to the quarry floor and between the vertical holes. The angle of the holes will vary with the terrain but they are rarely horizontal. They will normally be drilled at least one foot longer than the burden. Each hole is charged to not more than half its length and is closed with stemming material made up in cardboard cylinders. Toe holes are generally fired with delay detonators to provide a lifting effect after the main charge has shattered the bed rock.

Secondary drilling and blasting

This is the technique of reducing large slabs and boulders to a size which the primary crusher will accept. Initially such over-size rocks will probably litter the quarry site and will have to be cleared, probably forming a ground stock for early feeding to the crusher. As the face develops they will result from faults in the bedrock or from the pre-splitting effects of earlier blasts. A jackhammer will be used to drill such rocks and the hole is best drilled in the largest bulk and to a depth just beyond the centre, providing sufficient depth for adequate stemming. Experience will show the best charge for efficient splitting and minimum scatter and of course the charge will vary with the nature of the rock.

Plaster shooting is an alternative method of secondary blast-ing in which a charge of one or more explosive cartridges, primed with an electric detonator or with a detonator and safety fuse, is laid directly on a surface of the boulder. An alternative to the cartridges is Plaster Gelatine. The charge is covered by paper and then by a layer of plastic clay well pressed by hand and better tamping of the clay is achieved by wetting the rock surface.

Plaster shooting requires more explosive than pop shooting but has the advantages that:
(a) No drilling is needed.
(b) The boulder or slab is split in situ with no scatter and hence less danger from flying fragments.

A further alternative for secondary breaking is drop-balling. This requires a crane to hoist and drop a steel mass on to the oversize rocks. The "ball" will weigh around two tons and can be custom made for the purpose or can utilise, say, old piling

25

hammers or crusher shafts. An old rubber tyre inserted between the hoist wire and the weight will reduce back-lash and protect the wire from damage.

Quarry face planning

As the optimum drilling pattern becomes settled it will be possible to divide the face into sections for safe economical progressive working.

The sequence will have to take account of the number of benches, if any, but on a "straight" face i.e. with no fully formed benches the sections will be:

(1) A length being tidied up and cleared of loose rock ready for primary drilling.
(2) A length drilled with a pattern of holes being loaded for blasting.
(3) A length which has been fully blasted or is being drop-balled and loaded to the primary crusher.

This will be the sequence:

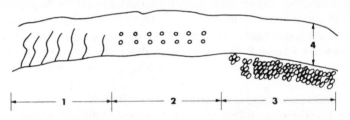

1. A LENGTH BEING TIDIED UP
2. A LENGTH DRILLED WITH A PATTERN OF HOLES
3. A LENGTH WHICH HAS BEEN FULLY BLASTED

Fig. 3

Of necessity all drilling and blasting operations will be governed by your local terrain and will need constant supervision and direction particularly if the operations are sub-contracted. The aim should always be a controlled progression towards the orderly and safe working of an established rock face on which the successive functions can be more-or-less tightly grouped.

This will minimise the disruption which is inevitably involved in withdrawing plant and personnel from the blasting area and will allow the loading and transportation of broken rock to the crusher to go ahead in a regular and safe manner.

Internal transport is considered in the next section of this manual.

Transport between the face and the crusher
The main choice for loading blasted rock at the face is between track mounted face shovels and wheeled front – end loaders. There are pros and cons for each method and points for consideration include:

(a) Initial capital cost – higher for face shovels.

(b) A face shovel can quickly be adapted for drop-balling.

(c) A wheeled loader can double up for loading out aggregates from the stockpiles.

(d) With a face shovel, bucket size must be matched to the feed opening of the primary crusher.

(e) When a wheeled loader is used care is needed to avoid loading rocks which are too large for the crusher.

(f) Wheeled loaders are susceptible to tyre damage on the quarry floor even when rock chains are fitted.

(g) Wheeled loaders are useful for the internal transportation of small maintenance items such as gas cylinders, compressors, spare parts etc. They can also more conveniently be used to provide lifts for re-metalling crushers etc.

(h) The loading cycle is generally quicker with a face shovel but the truck must be much nearer the pile of rock and thus more prone to tyre damage.

(i) A wheeled loader can traverse more quickly along the quarry face and can more readily be withdrawn to a safe distance during blasting.

(j) A wheeled loader is more versatile as an on-site towing vehicle.

The rock from the quarry is generally transported to the primary crusher in dump trucks although a wheeled loader can be used for small operations if the hauling distance is not significantly more than the minimum safe distance between face and plant i.e. around 250 metres. For continuous operation and to allow for maintenance there should be a minimum of two dump trucks and their capacity will be in the range of 5 – 10 tons. A reinforced body on a standard truck chassis is usual and there must be an extended apron over the cab to protect the driver who must stay in the cab during loading.

If standard highway trucks are used the floor of the body

should be reinforced by steel bars and the cab strengthened.

In all cases the access road between face and plant should have the easiest gradient permitted by the terrain. Tight curves should be avoided and the surface should be kept graded and free of loose rocks. Where the road traverses the edge of a drop large rocks should be placed at intervals along the edge to keep the trucks away from the drop.

At the feed point to the primary crusher a ramp and retaining wall will usually be needed.

For economical operation of the dump trucks make the ramp gradient as easy as the terrain will allow. Keep the ramp as straight as possible; if curves are essential use as large a radius as possible and make rock or soil banks on each side.

At the top of the ramp create an area large enough for the driver to reverse to the dump hopper by turning at full lock twice only. Also provide safe, flat ground for the plant attendants and a safe parking area, clear of the dump trucks, for maintenance vehicles, welding plant etc. If you employ more than one dump truck either (a) provide a passing place on the ramp or (b) be sure that the bottom of the ramp can be seen from the top.

Make provision for lighting columns to illuminate (a) the top of the ramp and (b) the dump hopper and the feeder.

The retaining wall is best built of cast concrete but jointed and bedded stone masonry can be used. In some areas and particularly for temporary sites spiked logs are employed. In all cases make the retaining wall wide enough at the top of the ramp to provide a flat area large enough for the safe and economical manoeuvering of the dump trucks. At the top of the wall incorporate either a steel dump hopper or a concrete chute leading into the feed hopper of the primary plant and install wing plates to eliminate over-spill. Make provision for a curtain of heavy chain links or old track pads to control the drop of rock into the feeder. At the upper end of the dump hopper or chute form a stop block for the wheels of the trucks. The wings of the retaining wall must be long enough to ensure that the filling material behind them does not spill onto or around the primary plant and will normally follow the gradient of the ramp almost to floor level.

The plant area
Most mobile or skid-mounted processing plants will comprise

one, two or three separate units linked by mobile conveyors for stockpiling sized aggregates. There may also be a mobile conveyor installed at an angle from the primary plant to take out quarry dirt from a grizzly before the primary crusher. There must be a flat and compacted area sufficiently large to accommodate these requirements and the plant manufacturer's arrangement drawings should be used to estimate the minimum area. Take care not to underestimate the space needed for stockpiles and bear in mind the necessity for operating a loading shovel and trucks within that area.

It is advisable to make a concrete raft for each major plant unit but concreting requirements can only be determined by reference to the local ground bearing capacity. *Consult a competent authority* and refer to your plant arrangement drawings for the point loads for each unit. It is advisable to make the concrete rafts approximately one metre wider and longer than the bearing points of the plant unit. This will help in keeping the ground under the plant clear of spillage and ensure safe footing and access. If in situ concrete plinths are to be used at the plant support stools in place of timber baulks one method is to use shuttering to form holes approximately 0.75 m square at each bearing point. Fill these with sand before locating the plant on the raft, then remove the sand and use the holes to key the plinths into the raft. Cast each plinth to a height within 2 cm from the stool sole-plate, insert steel packers and grout the stool.

Where timber baulks are used to support the plant use timbers of the largest available cross-section, properly squared, and aim for the minimum interfaces. For preference spike the timbers together.

THE PROCESSING PLANT AND PROCESS MACHINES

Modern mobile or skid-mounted aggregate production plants are unitised, that is to say each section, primary, secondary and tertiary, comprises a crusher with a feeder (primary) or a crusher with a screen (secondary and tertiary). There may be a final sizing unit comprising simply a screen mounted on its own chassis.

The specification and selection of the plant in terms of capacity, product sizes etc. are not within the scope of this

manual; it is assumed that a choice related to local requirements and availability has been made and will have taken
account of such factors as:

(a) Local availability of spares, particularly consumables.
(b) Facilities provided by the manufacturer or agent for training your personnel.
(c) The degree of accessibility provided on the plant units for
ease of operation and particularly maintenance.
(d) The extent of component standardisation between the
plant items.
(e) The suitability of the machines, particularly the crushers,
for your stone.
(f) The product flexibility available in the plant arrangement.
(g) The degree of portability related to your requirements.

In the following section we consider the types of individual
machines now preferred and likely to be incorporated.

Feeders

The first machine will be a feeder, the purpose of which is to
convey the rock from the dump hopper in a controllable flow
to match the capacity of the primary crusher. For hard rock
processing there are two main types:

(1) The apron feeder is continuous, rather like a belt conveyor,
and is well suited for handling large sizes of rocks, and
gravel. The overlapping steel flights, or pans, provide a self
cleaning action and the machine is well suited to handle
clay contaminated rock. It ensures a continuous positive
discharge of stone to the crusher but is at a disadvantage
with wet material since the feeder deck will leak. There is
no "grizzly" action i.e. dirt and undersize stones are not
removed from the feeder deck and it is usually necessary to
install stationary grizzly bars for this purpose between the
feeder and the crusher. If the undersize stone is clean it will
be by-passed via a chute to join the crushed stone under the
primary crusher. If there is contamination it will probably
be by-passed to a side conveyor for stock piling as quarry
dirt. The apron feeder is usually fitted with a device to vary
the speed of the deck and hence the feed rate.

(2) The mechanical vibrating feeder is the most popular feeder
now in use. The material is conveyed evenly forward by
vibrations at 45° to the horizontal and the deck usually

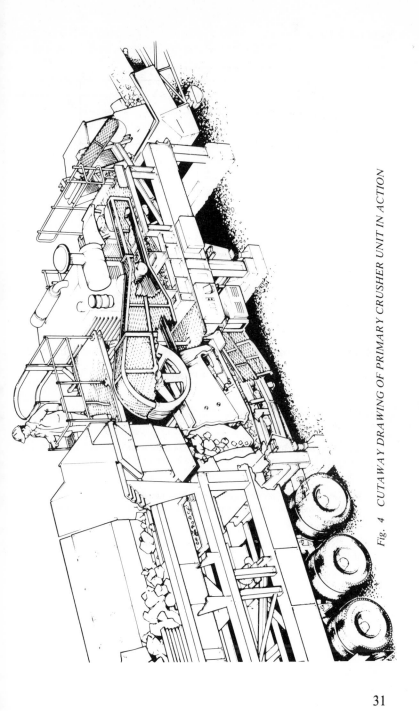

Fig. 4 *CUTAWAY DRAWING OF PRIMARY CRUSHER UNIT IN ACTION*

31

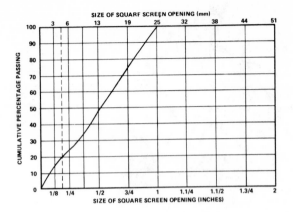

Fig. 5

Estimated plant capacity: 60 T.P.H.
Producing minus 25 mm (1″) material in two crushing stages.

Primary crusher:
Pegson-Telsmith 500 mm × 900 mm (20″ × 36″)
S.T.R.B. Jaw Crusher.
63 mm (2.1/2″) closed side setting, 60 T.P.H. capacity.

Secondary crusher:
Pegson-Telsmith 36 S(36″) Gyrasphere
14 mm (9/16″) closed side setting, 42 T.P.H. capacity.

Secondary crusher in closed circuit, with a 1524 mm × 4877 mm (5′0″ × 16′0″)
T.D. horizontal screen, separating product into 25 mm, 19 mm, 9 mm and 5 mm (1″,
3/4″, 3/8″ and 3/16″) sizes.

The above information is an average estimate based on crushing clean dry limestone
and will vary due to type and nature of material and operating conditions within about
± 5%.

For special circumstances, or for screening sizes other than indicated, please refer to
manufacturer.

(Figures 4, 5, 6 and 7 are reproduced by courtesy of Pegson Ltd.)

32

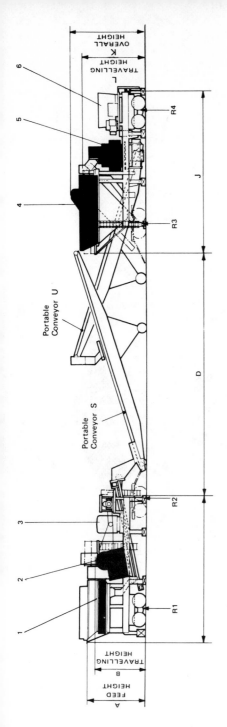

PRIMARY SECTION
1. VARIABLE SPEED VIBRATING GRIZZLY FEEDER.
2. SINGLE TOGGLE ROLLER BEARING JAW CRUSHER,
   AVAILABLE WITH AUTOMATIC HYDRAULIC
   ADJUSTMENT AND OIL LUBRICATION SYSTEM.
3. DIESEL DRIVE WITH OPTIONAL PLANT-MOUNTED
   GENERATOR OR STANDARD ELECTRIC MOTOR DRIVE.

SECONDARY SECTION
4. TRIPLE DECK HORIZONTAL VIBRATING SCREEN.
5. GYRASPHERE OR AUTOCONE SECONDARY UNIT.
6. DIESEL DRIVE WITH OPTIONAL PLANT-MOUNTED
   GENERATOR OR STANDARD ELECTRIC MOTOR DRIVE.

Fig. 6   TWO UNIT PORTABLE PLANT SPECIFICATION

33

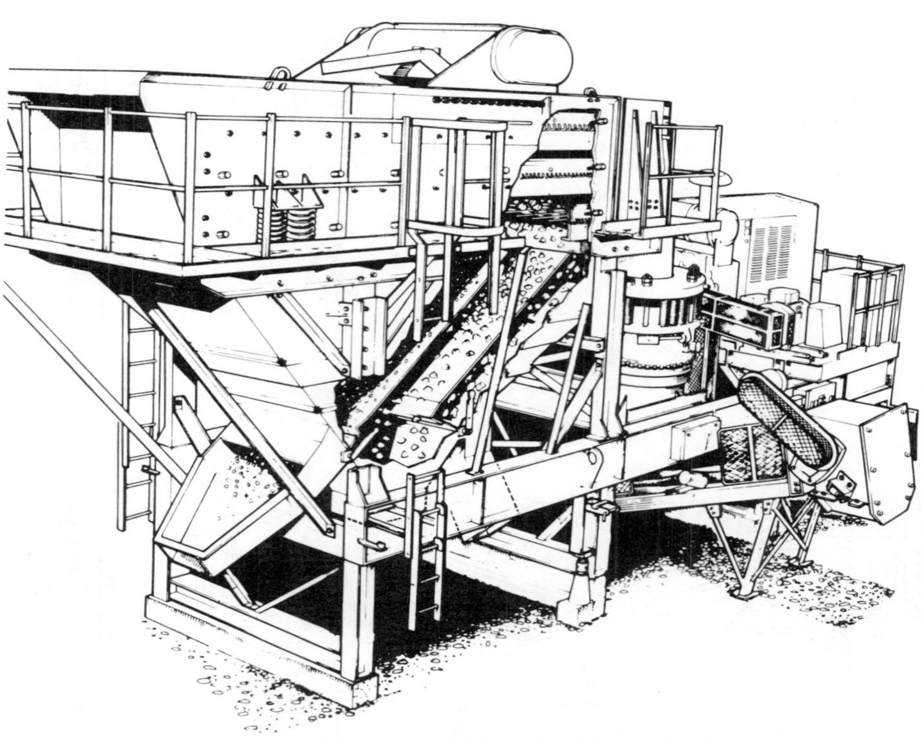

Fig. 7 A TYPICAL SECONDARY CRUSHER

incorporates spaced grizzly bars which vibrate with the plain pan. The bars can be stepped into two levels; the step will turn the stone and help to remove adhering fines and clay. The bars are adjustable for gap width and are tapered towards the discharge end. The speed and hence the frequency of the vibrations is usually controllable; preferably this should be possible by a multi-step motor control switch on the operator's platform. The bottom and sides of the pan should incorporate renewable steel liners.

Primary jaw crushers

The primary crusher follows the feeder and is almost always mounted on the same chassis together with a plant mounted discharge conveyor. The word "primary" does not indicate any specific design of crusher; it simply indicates the first crusher in the sequence.

On mobile plants jaw crushers are preferred. The first generation of jaw crushers were "double toggle" machines in which an eccentric shaft activates a pitman which in turn raises and lowers two toggle plates rocking the jawstock. These are also known as "Blake" type crushers and since there is more movement at the bottom of the jawstock than at the top there is no force feed action. This means that the crusher is somewhat unsuitable for clay contaminated rock. It is also a heavy, expensive machine to produce and for mobile plants single toggle jaw crushers are more often specified. On these crushers the jawstock itself is hung directly on and is activated by the eccentric shaft. There is a force feed action which makes the single toggle crusher an efficient machine on sticky rock.

There are variations of toggle plate angle and in the method of adjusting the discharge setting but the operating principle is usually common regardless of the origin of the crusher. The reduction ratio of a jaw crusher is normally around 7:1.

Secondary and tertiary crushers

Secondary and tertiary crushers are of two main types:
(1) Jaw granulators, generally single toggle type similar in action to the single toggle primary crusher. These are only suitable for comparatively low output operations and on hard rock are subjected to heavy wear necessitating frequent replacement of the fixed and swing jaw plates.

(2) Cone crushers are more commonly specified, being capable of high outputs of uniformly crushed aggregates. On mobile plants potential outputs range from approximately 4t/h to approximately 400t/h and styles are available for very fine crushing (not normally used on mobile plants)

1. RENEWABLE SIDE LINERS
2. PLAIN PAN WITH RENEWABLE LINERS
3. TAPERED GRIZZLY BARS WITH VARIABLE SPACING
4. SPRINGS

Fig. 8   A VIBRATING GRIZZLY FEEDER

Photo courtesy Pegson Ltd

1. DRIVE GEARS TO HEADSHAFT
2. TOP SUPPORT ROLLERS
3. LINKED APRON PLATES
4. CENTRALISED LUBRICATION
5. RETURN SUPPORT ROLLERS

Fig. 9   A PLATE APRON FEEDER

Photo courtesy Pegson Ltd

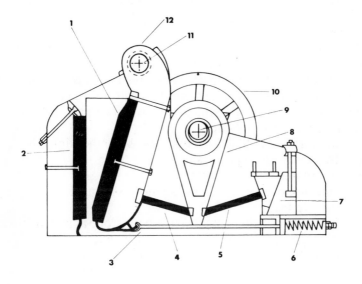

1. SWING JAW
2. FIXED JAW
3. TENSION ROD
4. TOGGLE PLATES
5. TOGGLE PLATES
6. SPRING
7. ADJUSTMENT
8. PITMAN
9. ECCENTRIC SHAFT
10. FLYWHEEL
11. JAWSTOCK SHAFT
12. JAWSTOCK

Fig. 10   CROSS SECTION OF A DOUBLE TOGGLE JAW CRUSHER

fine, standard, coarse and extra coarse operation. The reduction ratio of a cone crusher is around 10:1.

Traditionally cone crushers have been equipped with manual adjustment and spring relief for overload; they are available now with rapid hydraulic adjustment and automatic overload release and resetting.

(3) Roll crushers are occasionally used for secondary and tertiary reduction but are not generally favoured. The maximum feed size is directly related to the diameter of the crushing rolls and must not exceed 3 times the roll gap otherwise the stone will slip and induce heavy localised wear. Localised wear will also result if the feed is not distributed across the full face width of the rolls. Roll crushers are not really suitable for use on abrasive stone which will often necessitate daily weld rebuilding of the rolls. The reduction ratio of a roll crusher is no more than 3:1.

Screens

A modern secondary or tertiary plant unit will usually incorporate a screen as well as the crusher. Apart from the cylindrical rotary screen used on small single-unit plants the screen will have one or more flat decks, mechanically vibrated and fitted with perforated media which can be wire mesh, perforated plate or moulded polymer.

The vibration is generated by one of several options such as:
(a) One or more eccentric shafts.
(b) A counterweighted shaft.
(c) One or more electric vibrating motors mounted directly on the live frame.

There may be water sprays to rinse the stone as it passes over the decks. Some vibrating screens are inclined, in which case there will be one shaft or electric vibrators.

Horizontal screens require less head room and are thus favoured for mobile and skid-mounted plants. There will be two shafts coupled by gears and providing a vibrating action very similar to the method employed on vibrating feeders as already described.

Conveyors

In addition to the main processing machines the plant complex will include belt conveyors to transport the stone between the process stages and to stock-pile the finished aggregates. Some of the conveyors will be plant-mounted i.e. they will be built into the unit structure. Others may be short, free-standing conveyors. There will also be wheel-mounted conveyors, perhaps fitted with swivel axles for radial operation and with adjustable discharge height. Lengths and belt widths vary and will be related to the overall plant requirements.

On comparatively small installations, up to say 50 tonnes per hour, it may be possible to dispense with some of the mechanical conveyors and use manual porterage instead. Wheeled barrows or head-baskets can be considered but take care to provide firm, level pathways and secure ramps for the carriers.

When porterage is employed make sure that any revolving machinery near which the people must pass is adequately guarded.

Drives
The plant will incorporate a variety of drives which will mainly comprise:
(a) Vee belts and grooved pulleys.
(b) Gearboxes which can be long coupled to electric motors, shaft-mounted on electric motors, or mounted directly on the driven shaft e.g. a conveyor head shaft.
(c) Roller chains and sprockets.
(d) Fluid couplings as occasionally used between a power unit and its driven machine.

Power units will be diesel engines, electric motors or a combination of both. Where electric motors are used the current source might be:
(a) In the case of all-electric drives a separate diesel generator set or mains supply.
(b) In the case of combination drives, a plant-mounted alternator, probably around 75 KVA and driven by a plant mounted diesel engine. This engine will also directly drive the crusher; the alternator will provide current for the other drives e.g. feeder, screen, conveyors, together with current for plant lighting, power tools etc.

Metal detectors
A metal detector is a useful adjunct. "Tramp" metal such as

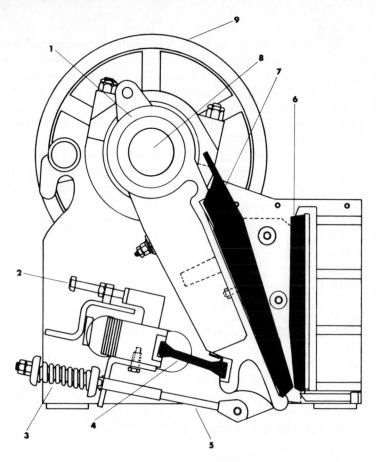

1. JAWSTOCK
2. ADJUSTMENT
3. SPRING
4. TOGGLE PLATE
5. TENSION ROD

6. FIXED JAW
7. SWING JAW AND WEDGE
8. ECCENTRIC SHAFT
9. FLYWHEEL

Fig. 11   CROSS SECTION OF A SINGLE TOGGLE JAW CRUSHER

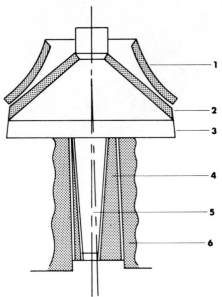

1. BOWL LINER   3. CONE HEAD   5. MAINSHAFT
2. HEAD LINER   4. ECCENTRIC   6. MAIN FRAME

Fig. 12   OPERATING PRINCIPLE OF A CONE CRUSHER

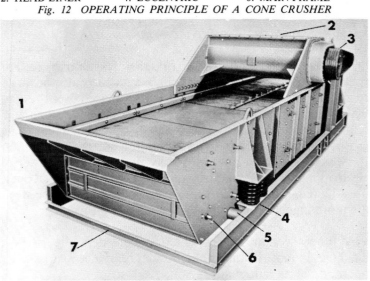

1. FEED END   5. SNUBBER SPRING
2. VIBRATING MECHANISM   6. MESH FIXING BOLTS
3. DRIVE PULLEY   7. SUPPORT FRAME
4. SUPPORT SPRINGS

Fig. 13   A HORIZONTAL SCREEN

Photo courtesy Pegson Ltd

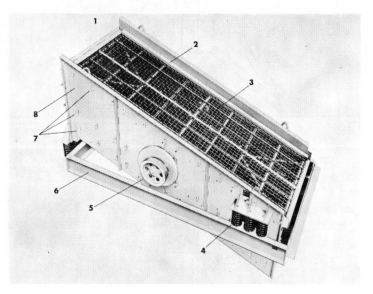

1. FEED END
2. MESH TENSIONING PLATES
3. WIRE MESH AND SUPPORTS
4. SPRINGS
5. DRIVEN PULLEY
6. SUPPORT FRAME
7. MESH FIXING BOLTS
8. LIVE FRAME

Fig. 14 A TYPICAL INCLINED SCREEN

Photo courtesy Pegson Ltd

drill bits and hammer heads can find their way into the feed of rock to the primary crusher although every care must be taken to avoid this possibility. A metal detector placed over the conveyor between the primary and secondary crushing stages will minimise the chances of uncrushable metal damaging these crushers or at best delaying production while the obstruction is cleared. The metal detector may incorporate a powerful electro-magnet to lift tramp metal from the conveyor belt but it must be remembered that some metals, e.g. manganese steel, are non-magnetic and cannot be removed by this method.

Another section of this manual considers the process machines in more detail with particular reference to aspects of operation and maintenance.

42

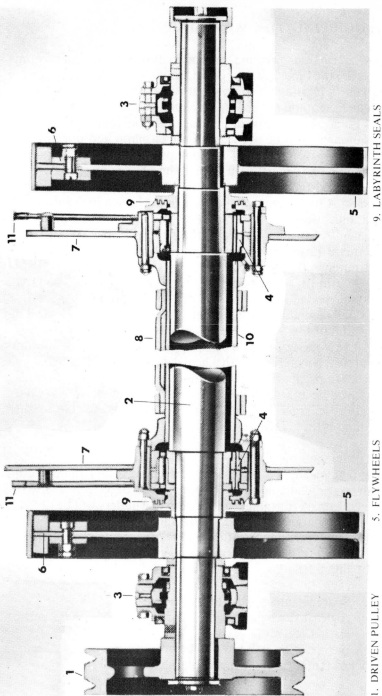

1. DRIVEN PULLEY
2. ECCENTRIC SHAFT
3. OUTER ROLLER BEARINGS
4. INNER BEARINGS

5. FLYWHEELS
6. VARIABLE COUNTERWEIGHTS
7. SCREEN LIVE FRAME
8. RUBBER PROTECTIVE COVER FOR CROSS TUBE

9. LABYRINTH SEALS
10. CROSS TUBE
11. GREASE NIPPLES AND TUBES

Fig. 15 SHAFT ASSEMBLY OF A TYPICAL INCLINED SCREEN

EQUIPPING THE WORKSHOP

At an early stage in the development of the site and certainly before the time comes to position and assemble the processing plant, a reasonably comprehensive selection of general maintenance equipment should be available.

For the loaders and dump trucks normal road vehicle garage equipment will be needed and diesel pump and injector calibrators will be useful. There should also be a compressor capable of 10 to 30 c.f.m. at maximum working pressure 150 p.s.i. and tyre changing equipment matched to your largest tyres. A hydraulic trolley jack and a wheeled hydraulic crane should also be considered. For greasing vehicles and plant generally a manually operated high pressure drum pump with a 2 metre hose and interchangeable connectors will meet the requirements of the comparatively small workshop.

A grinder will be needed, preferably pedestal mounted, double ended with wheels of minimum diameter 200 mm. A second special purpose grinder should be installed for drill bit grinding.

The workshop should be equipped with at least one substantial steel bench fitted with a heavy duty vice of at least 200 mm × 100 mm capacity. A stand-mounted tube vice, 75 mm capacity, will be useful, as also will a heavy steel anvil. A heavy duty battery charger will be needed. It may be possible to sub-contract these facilities.

Cutting and burning

A regular supply of gas must be ensured and there should always be available on site three cylinders of oxygen and two of acetylene. A portable trolley for two cylinders will be needed.

The equipment should also include:
(a) Two spare gauges of each type (oxygen & acetylene).
(b) Two cutting torches with a selection of nozzles.
(c) Spare lengths of oxygen and acetylene hoses, say 20 m of each.
(d) Spare spanners, keys and nozzle cleaners.
(e) At least three goggles, four pairs of gloves and a pair of leather leggings.

Consideration will have to be given to providing a secure, locked store for all of this equipment.

Welding

A portable welding set is essential and there is a wide choice. The set should have an output of 300 to 400 amps and must be of rugged construction, the oil cooled transformer type being generally preferred. An extension cable of at least 25 m will be needed.

An additional engine-driven, truck-mounted set might be considered.

The equipment should include:

(a) Standard duty low-hydrogen rods for vertical and horizontal welding in gauges 6, 8, 10.

(b) Hard facing rods in gauges 6 & 8.

(c) At least four pairs of gloves, one hand held screen and one head screen, both with spare glasses.

(d) Chipping hammers and wire brushes.

(e) Heavy duty steel "G" cramps, size 250 mm

(f) Rack cramps, size 300 mm

Portable lifting equipment

This requirement is important and should be adequately covered to discourage, for example, the use of old vee belts as slings.

The following should be considered.:

(a) Wire rope slings, 2 – leg or 4 – leg, lengths 3 m, for 2 ton and 5 ton loads.

(b) Chain slings, single and double, length 3 m, for 2 ton and 5 ton loads.

(c) A 3 ton chain type pull lift.

(d) Two ratchet screw jacks, 10 ton rating.

(e) Two double ram hydraulic jacks, 10 ton capacity.

(f) A selection of screw pin "D" shackles, 1 ton to 5 ton rating.

(g) A selection of standard lifting eye bolts in screw diameters from 12 mm to 25 mm.

Plant erection tools

In addition to the general purpose lifting equipment there will be special requirements for erecting and assembling the plant. These will include:

(a) Crow bars and pinch bars.

(b) 1 kg ball pein hammers and 6 kg sledge hammers.

(c) Cold chisels.

(d) Podgers and drifts.

(e) A 0.5 m spirit level.

(f) A 1 m square

(g) A set of conveyor belt fastener tools i.e. belt punch, bolt wrench, bolt breaker and template.

(h) A plumb bob.

(i) A string line.

(j) A wind up measuring tape, 50 metres.

(k) A selection of flat, round and half round files.

(l) Metric allen keys to 26 mm and Imperial allen keys to 1″.

(m) Open end or ring wrenches to metric 50 mm.

(n) Open end or ring wrenches to Imperial 2″.

(o) A supply of wax chalk.

A 6 m extending ladder will also be useful.

General tools

A lockable steel box should be kept in the workshop to contain the small tools from the above list and:

(a) 20 cm – 30 cm adjustable hack-saws and spare blades.

(b) A 70 cm rip-saw.

(c) 30 cm & 15 cm engineer's screw drivers.

(d) 25 cm & 20 cm adjustable spanners.

(e) A 10 cm chainpipe wrench.

(f) 30 cm & 60 cm Stillson wrenches.

(g) Combination pliers and heavy duty electric pliers.

(h) Open end or ring wrenches up to metric 26 mm and Imperial 1″.

(i) An Avometer with carrying case.

(j) A clipper or tongue tester rated to 600 volts and 300 amps.

A heavy duty portable electric drill will be needed and should be rated at 110 volts for use with a portable tool transformer with an adequate extension cable, say 30 m. The drill should have a minimum capacity in steel of 12 mm and must be fitted with a side handle. There must be an adequate supply of straight shank high speed twist drills up to size 12 mm.

There should also be available in the workshop or in the office a comprehensive first-aid kit which should contain telephone numbers and addresses of the nearest outside medical facilities.

SAFETY IN QUARRIES

There will almost certainly be federal and perhaps local regulations governing safety, health and welfare and you will have an unalienable responsibility to aquaint yourself with and comply with these requirements. It will probably be mandatory to exhibit certain notices of statutory obligations.

The following points represent a common-sense approach to safety but in no way qualify the legal responsibilities:

(1) The use of protective equipment such as safety helmets, safety boots, goggles and gloves should be consistently encouraged, with the management and supervisory staff taking care to set a good example.

(2) Blasting notices should be displayed. There must be strong audible warnings of the start and end of blasting and there should be adequate blasting shelters, remote from the rock face but within easy reach of all personnel.

(3) Adequate ladders and railed platforms must be provided and local regulations will indicate the minimum height at which these are mandatory.

(4) There should be accessible emergency stop switches adjacent to each electrically powered machine. This is particularly important in the case of belt conveyors.

(5) The management must insist that all drive guards are maintained in good condition and are replaced immediately after maintenance work. The supplier may have no responsibility in law for adequate guarding; this responsibility rests entirely with the owner or his agent.

(6) The owner or his agent is also responsible for the safe operation of all activity at the quarry face, particularly in guarding against injury or damage by rock fall. There must be regular inspections and all personnel should be alert to the possibility of loose boulders or fragments falling from the face. The inspections should take particular notice of the possibility of falls resulting from fissures or bedding planes or from pre-splitting caused by earlier blasts.

(7) Points for particular attention of personnel working on or around the processing plant are:
   (a) Follow closely the instructions of your manual(s) whenever maintenance work is undertaken.
   (b) Always observe the regulations appropriate to your

area concerning the safe use of lifting chains and slings, ladders, platforms etc.

(c) Before commencing any maintenance work isolate the electric motors and control panels. Switch off at the isolators, remove the fuses and warn any other personnel involved. Do not start work until all drives are stationary.

(d) Make sure that all guards are replaced and secure before restarting.

(e) Check also that all personnel are standing clear of drives and moving machinery whenever any machine on the plant is started either at the beginning of the shift or after maintenance work.

(f) Always wear a safety helmet when working on or near the plant and avoid the use of loose clothing if your work takes you near moving machinery.

(g) Maintain good "housekeeping" on and around the plant:

   (i) Clear up any spillage, particularly on platforms and seal off any spillage points as they develop.

   (ii) Do not allow any scrap components, old vee belts, worn conveyor rollers, discarded maganese steel, timber etc. to lie on or around the plant. *Take it all to a scrap heap clear of the working area.*

   (iii) Extend good "housekeeping" to the quarry floor. Do not allow digger teeth, drill bits etc. to reach the crushing plant with the stone.

(h) Use platforms and ladders. Do not take short cuts by climbing steel-work.

(i) When using lifting devices make sure they are securely anchored. Before an object is lifted, make sure it is firmly and safely held.

(j) When working above ground level make sure that your tools cannot fall on people below you.

(k) Never stand or work under a suspended load or men working above you.

(l) Never interfere with electrical switchgear — get an electrician.

(m) Do not allow dust and spillage to accumulate around electrical control equipment or in any way restrict the air flow around control panels and resistor banks.

(n) Store oily rags in containers.

(o) Do not store inflammable liquids near an electric motor, control panel or engine.

Remember : *Safety warnings do not by themselves eliminate danger. The instructions or warnings they give are not substitutes for proper accident prevention measures.*

PART TWO: GENERAL MAINTENANCE AND OPERATION

In this section we consider the day by day maintenance and operation of the main machines and ancillary equipment brought together in a mobile or skid mounted aggregate processing plant beginning with some requirements for preparing the plant for duty.

Preparing the plant for duty

It will of course first be necessary to locate the plant units on their concrete pads or on level, hard ground, ensuring that there is a flat area large enough for manoeuvering the wheeled conveyors into position. Any components which have been transported separately must then be reassembled in their operating form.

A unit on road wheels will be provided with jacks to take the load off the running gear. Timber baulks will be placed under the jacks and the jacks operated in sequence to take the load off the wheels, taking care to keep the plant level across and along the chassis.

There will usually be separate support stools which are now bolted into place over concrete plinths or heavy timbers and the jacks lowered in turn to transfer the weight of the plant to the stools. A spirit level should be used to check the level in each direction and if necessary packers inserted beneath the stool sole plates. Now ensure that all oil-lubricated units are filled to the correct level with the recommended oil.

Next check that all vee belt and chain drives are correctly aligned and tensioned and that all drive guards are in position clear of the moving parts.

Conveyor belts will probably need to be properly tensioned and tracked when power is available for running the conveyors and at this time the belt scrapers can be adjusted, being particularly sure that they are not fouled by the belt fasteners.

Electrical connections will have to be made as detailed on the plant manufacturer's schematic drawings and earthing requirements seen to. Ensure that all cables, particularly on mobile conveyors, are clear of moving machinery and conveyor belts.

If diesel engines are incorporated attend to:

(a) The service tank fuel supply.

(b) Lubricating oil levels.

(c) Battery charging and connections.

(d) Air filter elements and pipes or hoses.

(e) Coolant level and anti-freeze if appropriate.

(f) Satisfactory clutch operation.

(g) Clearance around the fan and full access of air to the radiator.

Where electric motors are fitted check the free running of the fan and free flow of air.

Locate all grease nipples and give each two or three shots of the recommended grease.

Lubricants suited to most conventional drives are mentioned on page 96.

Always check your manuals for the manufacturer's own recommendations.

If your plant incorporates electrical control panels check that all fuses are intact and correctly rated and that there is free flow of air around the cabinets. There may be electrical timers which must be properly adjusted and sequenced, referring to the wiring diagrams for settings. If there are liquid immersed starters, fill the tanks with the specified electrolyte to the indicated levels.

It will normally be advisable to remove road running gear, where fitted, for secure storage elsewhere but if the wheels with tyres and hubs are to be left in situ cover them for protection against dust, accidental damage and direct sunlight. Put the hand parking gear into the "on" position and grease the wire cables and operating screws.

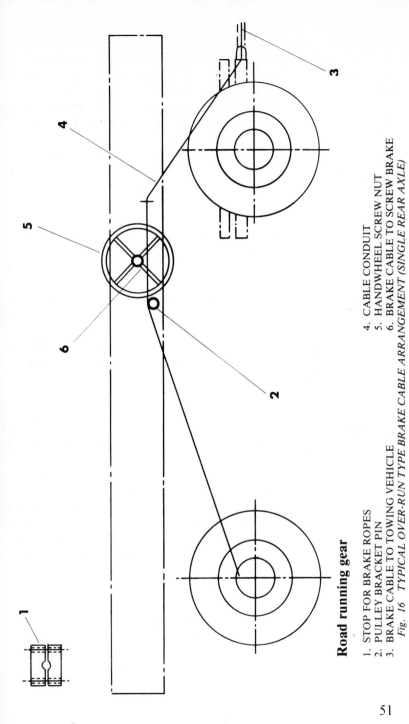

Road running gear

1. STOP FOR BRAKE ROPES
2. PULLEY BRACKET PIN
3. BRAKE CABLE TO TOWING VEHICLE
4. CABLE CONDUIT
5. HANDWHEEL SCREW NUT
6. BRAKE CABLE TO SCREW BRAKE

Fig. 16 TYPICAL OVER-RUN TYPE BRAKE CABLE ARRANGEMENT (SINGLE REAR AXLE)

51

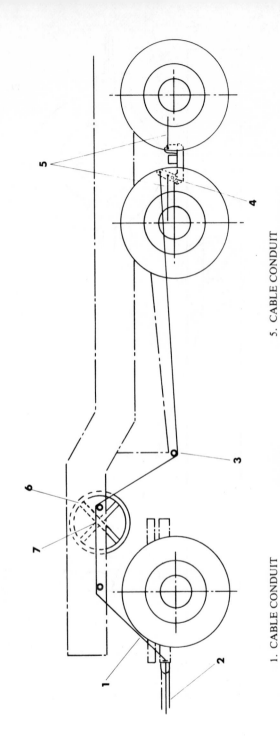

1. CABLE CONDUIT
2. BRAKE CABLE TO TOWING VEHICLE
3. PULLEY BRACKET PIN
4. PIN

5. CABLE CONDUIT
6. HANDWHEEL SCREW NUT
7. BRAKE CABLE TO SCREW BRAKE

Fig. 17   TYPICAL BRAKE CABLE ARRANGEMENT (DOUBLE REAR AXLE)

52

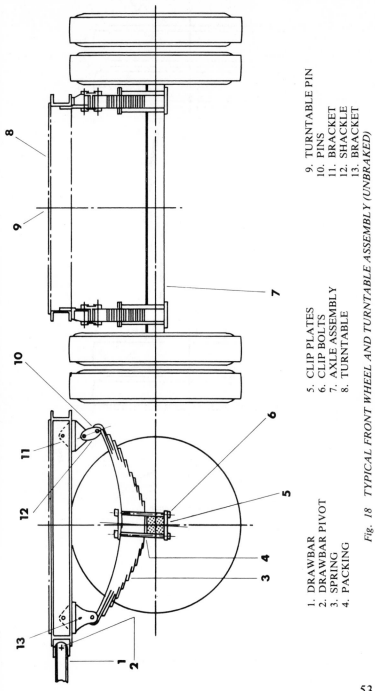

1. DRAWBAR
2. DRAWBAR PIVOT
3. SPRING
4. PACKING

5. CLIP PLATES
6. CLIP BOLTS
7. AXLE ASSEMBLY
8. TURNTABLE

9. TURNTABLE PIN
10. PINS
11. BRACKET
12. SHACKLE
13. BRACKET

Fig. 18   TYPICAL FRONT WHEEL AND TURNTABLE ASSEMBLY (UNBRAKED)

53

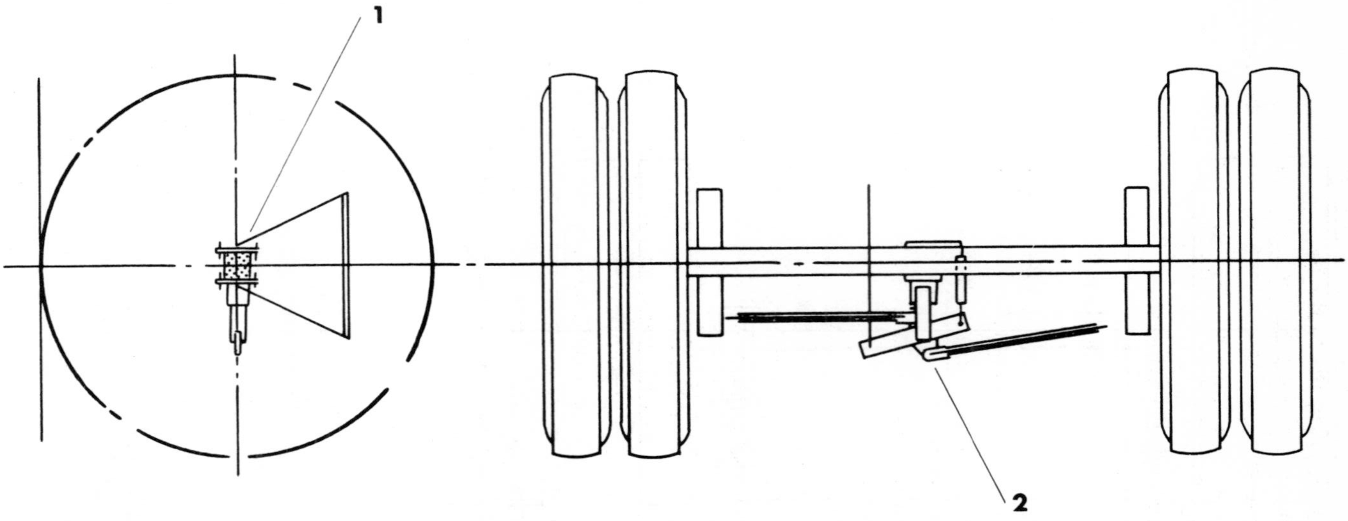

1. HUB ASSEMBLY
2. BRAKE GEAR

Fig. 19 TYPICAL REAR WHEEL ASSEMBLY (SINGLE AXLE) WITH CABLE BRAKES

54

Moving a plant on the highway

Before moving a plant on the highway:

(1) Reduce the overall height and width by detaching such items as are specified by the manufacturer. These may include:
  (a) Feeder and hopper.
  (b) Vibrating screen.
  (c) Ladders and platforms.

(2) Check:
  (a) Air lines and brake chambers.
  (b) Hubs and brakes.
  (c) Tyres and tyre pressures.
  (d) Brake lights.
  (e) Side marker lights.
  (f) Rear lights.
  (g) Registration plate.
  (h) "Towing" triangles.

(3) Carefully inspect the complete plant to ensure that there are no loose parts which may become detached during towing.

(4) Be certain that the driver of the towing vehicle is aware of the height, width, length and ground clearance of the plant unit. He must also be advised of the gross weight and the proposed route should be surveyed with all of these factors in mind.

(5) Notify those authorities, particularly the police, who may require knowledge of the movement.

Movement of the larger plants for short distances on the quarry site maybe permissible with the plant fully assembled provided that the route is firm and level. Normally however the speed should not exceed 1.2 k.p.h. and the turning angle should not be more than 20°.

DRIVES AND TRANSMISSIONS

Your processing plant will certainly incorporate a variety of drives and transmissions to propel the feeder, crusher, screens and conveyors. These will probably include gearboxes, vee belts and roller chains and are as important to the satisfactory operation of the plant as the processing machines which they

55

drive. Many involuntary shut-downs result from lack of attention to drives and transmissions which are often taken for granted until a failure occurs.

The equipment will usually be of proprietary origin, "bought in" by the plant manufacturer.

When ordering replacement parts the following details will be needed by your supplier:

For gearboxes:
From the identification plate:
(a) Manufacturer and type reference.
(b) Serial number.
(c) Increase or reduction ratio or output speed.
(d) Input speed.
Plus the following in the case of gearboxes flange-mounted on electric motors.
(e) Motor frame size.
(f) Motor speed.

For vee belt drives
(a) The vee belt code reference which will indicate the cross-section size and the pitch circle length.
(b) The number of vee belts which comprise the matched set.

For chain drives
(a) The pitch of the chain.
(b) The chain form i.e. simple, duplex, triplex.
(c) The number of links, including the connecting link and possibly a cranked link.

It is advisable to keep in stock on site at least one matched set of vee belts for each drive and lengths of chain with connecting links for each type employed on the plant.

Vee belt drives
The satisfactory maintenance of vee belts and pulleys will help to achieve maximum output from your plant. In particular it is important to maintain correct vee belt tension, which is achieved by several means, depending on the nature of the drive, i.e.:
   (i) By the slide rails of an electric motor or the sliding base of an engine.

(ii) By the bolts and slotted holes of shaft bearings.

(iii) By the screwed tensioning rod of a torque-arm gear-box.

(iv) The belt of the variable speed unit of an apron feeder is a special case (see below):

(v) The drive of a vibrating grizzly feeder is another special case; on these machines belt tension is often maintained by a spring.

As a visual check, there should be only a slight bow on the slack side when the belts are under load.

Never use a pinch-bar or similar tool to force vee belts into their grooves; always reduce the centre distance of the drive, using the appropriate adjustment method as listed above, until the belts enter the grooves without forcing.

Immediately remove with petrol or white spirit any grease or oil which may contact the vee belts.

Do not allow vee belts to turn over or twist in their grooves. Replace any which eventually develop this tendency but wherever possible replace as a matched set.

Keep the pulleys in alignment and their shafts parallel; this is easily checked with a taut cord or a straight edge.

Replace pulleys when the grooves become worn or accidentally damaged.

Diesel engines

Many different types and makes of diesel engine are specified; the specification varying with the power requirements and customer preference etc.

These notes are therefore necessarily brief and of a general nature.

For fuller information and for parts lists refer to the manuals supplied with your engine.

Routine operating and maintenance points

1. Never start the engine unless the belt drive guards are in position.

2. Do not smoke near batteries; and keep them well ventilated.

3. Before working on the electrical system of an engine detach and tape the positive battery lead.

4. Do not operate an engine with the governor linkage disconnected.

57

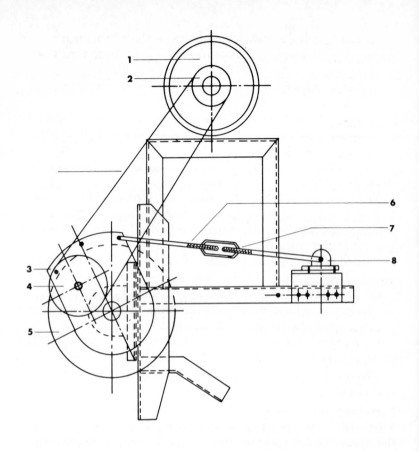

1. ELECTRIC MOTOR
2. DRIVE PULLEY
3. TORQUE ARM GEARBOX
4. DRIVEN PULLEY
5. CONVEYOR HEAD DRUM
6. TORQUE ARM ASSEMBLY
7. TURNBUCKLE
8. ANCHOR

Fig. 20   TYPICAL TORQUE ARM CONVEYOR DRIVING GEAR (SIDE ELEVATION)

Pre-start checks

5. If battery chargers have been in use − disconnect them.
6. Check the air cleaner. On larger engines there is usually an air cleaner service indicator. If the red piston is locked in the visible position, service the air cleaner as instructed in the engine manual.
7. Check the lubricating oil level on the dipstick. Add clean oil if necessary.
8. Check the coolant level in the radiator of a water cooled

engine and top up if necessary.
9. Disconnect the driven unit if a clutch is provided.
10. If the engine is driving a generating set put the load switch
into the "OFF" position.

After starting
11. If an oil pressure gauge is fitted, check that it is reading
correctly according to the manual.
12. After a few minutes similarly check the oil and water temperature gauges. *Stop the engine immediately if you are
not satisfied with any of these readings.*
13. Check for and rectify any leaks in the fuel, water and lubricating systems.
14. Check the dynamo charging ammeter where fitted.

Electric motors
These notes relate to electric motors generally. When you have
to order spare parts or refer to a particular motor always quote
the following details from the identification plate:
(a) Name of manufacturers.
(b) Frame size.
(c) Serial number.
(d) Current characteristics.
(e) Horsepower or kW rating.
(f) Speed.
(g) Driven machine. This is particularly important in the case
of motors flange-mounted on gearboxes, usually encountered on conveyors.

*If the driven machine is a horizontal or vibrating screen the
motor must generally be wound or rewound for 225% starting
torque*

Routine operating and maintenance points
1. Protect the motor against damp and falling moisture
(unless the motor is specifically protected against this).
2. Avoid the entry of dirt and fluff which may block the
ventilation.
3. Do not place anything on or around a motor which may
restrict air flow.
4. Motors should not be located in an area where the ambient

temperature exceeds 40°C unless specifically designed for this duty.

5. Motors fitted with grease fittings and relief plugs should be re-lubricated as follows:

(a) Wipe clean the pressure-gun fitting and the area around the motor grease fittings.

(b) Remove the relief plug and free the relief hole of any hardened grease.

(c) With the motor stationary add grease until new grease is expelled through the relief hole.

(d) Run the motor for about 10 minutes with the relief plug removed to expel surplus grease.

(e) Clean and replace the relief plug.

Shell Alvania R2 or R3 grease or an equivalent is recommended.

Motors with sealed bearings should not require regreasing.

Cleaning

6. Expose the windings by removing the endshields and the rotor. In the workshop it is preferable to remove the dust by vacuum cleaner. For in situ cleaning use a comparatively low pressure jet of air — a fierce jet will pack the dust further into the ventilation ducts and the interstices in the winding. Take care to avoid damaging the insulation.

7. Slip-ring motors having ventilated slip-ring covers should be examined periodically and any accumulation of dust on the insulation of the slip-rings and brush gear removed.

8. Take care to avoid oil or grease on the winding. Petrol, trichlorethylene or carbon tetrachloride may be used to remove oil or grease. These solvents must however be applied sparingly and must not come into contact with rubber cables, particularly the leads between the winding and the terminal block. When applying solvents take account of the inflammable or objectionable nature of the fumes.

Slip-ring maintenance

9. If the slip-rings are not badly worn they may be polished by holding a piece of finest grade carborundum cloth against the rings whilst they are rotated at speed, afterwards

removing all traces of carborundum and metal dust. Rough or grooved slip-rings should be removed and ground or turned and polished.

10. New brushes must be bedded to the curvature of the slip-rings. To do this place a strip of emery cloth between the slip-rings and brushes, the rough surface towards the brushes. Hold the brushes to the emery cloth by normal spring pressure and move the emery cloth under the surface of the brushes until the full width of each brush fits the surface of the slip-ring. Carefully remove all traces of carbon dust, particularly from between the brushes and brush holders. After bedding do not change the brushes from one brush holder to another or turn them round in the brush holders.

Suggested sequence for tracing faults in an electrical control panel

(1) Check that all fuses are intact and of the specified rating.
(2) Check that all timers and overloads are correctly adjusted.
(3) Disconnect all outgoing supply wires at the terminals.
(4) Switch the out-of-sequence switch to the OFF position.
(5) Turn the main switch to the ON position. There should be no response within the cabinet.
(6) Turn the main switch to the OFF position.
(7) Reconnect the drives one at a time, testing and proving each before reconnecting the next one until the faulty circuit is identified and reconnected.

THE MAIN PROCESSING MACHINES

These machines, feeders; primary crushers; secondary crushers; tertiary crushers and vibrating screens, will of course vary in type and detail with the origin of the plant complex. It is therefore absolutely essential to keep the manufacturer's own service manuals available on site and to refer to them for all aspects of operation and maintenance.

The notes which follow are of necessity somewhat general and should be regarded only as guides to be qualified and amplified by the appropriate service manuals.

Routine operating and maintenance points

(1) *Vibrating grizzly feeders and apron feeders:*

(a) Always disconnect the main power source to the feeder before starting any maintenance work.

(b) Do not operate the feeder unless the drive guards are in position.

(c) The presence in the feed of rocks too large to enter the mouth of the primary crusher will cause lengthy stoppages of the entire plant. Such rocks should be eliminated from the feed at the loading point on the quarry floor but if one reaches the dump hopper stop the feeder *before* the rock reaches the crusher. Remove the rock with a grapple if available or with a wire rope. Alternatively it may be possible to split the rock on the feeder deck – but *not* with explosives.

(d) If it is necessary for anyone to enter the dump hopper or the feed hopper for any reason, including (c) above, *switch off the feeder motor and the crusher motor. Station a man on the ramp to ensure that no dump truck tips until the hopper is clear of personnel.*

(e) On oil-lubricated vibrating grizzly feeders look at the oil level sight indicators at the start of each shift and top up the oil if necessary. If the indicators become discoloured fit new ones, wash the old ones in petrol and put them into store.

(f) On vibrating grizzly feeders maintain the gap between the pan side plates and the feed hopper skirt plates as nearly parallel as possible on both sides to eliminate trapped stones.

(g) Similarly on apron feeders maintain the overlap between the feed hopper skirt plates and the flight plate flanges equal and parallel on both sides.

(h) Replace any hopper liners, frame liners and pan liners as they wear.

(i) Particularly on vibrating feeders replace without delay any missing bolts and screws and pay particular attention to keeping the grizzly bar bolts tight.

(j) Replace immediately any broken or "tired" springs.

(k) On apron feeders keep all shafts, particularly the head and tail shafts, parallel to each other and square to the feeder frame.

(l)   At least once every shift screw down all grease cups and refill them if necessary. Similarly give any grease nipples one or two shots of grease, referring first to the service manual for recommended lubrication periods.

(m)   An apron feeder will usually incorporate a device to vary the speed of the deck. It may be advisable, after reference to the manufacturer's manual, to operate the variable speed drive daily through its full range to eliminate the possibility of seizure.

It is advisable, particularly in the case of vibrating feeders, to institute a preventive maintenance routine and the following is a typical schedule:

Daily:
(a) Check oil levels − both sides − and replenish if necessary.
(b) Check pivots of motor baseplate for free movement.
(c) Check that vee belts are correctly located in grooves of both pulleys.
(d) Check that there is no movement of grizzly bars − retighten or replace bolts as necessary.
(e) Check for loose or missing bolts in pan liners and frame liners.
(f) Check that the drive guards are in position and secure.

Weekly:
(a) Check the condition of pan liners and frame liners. Replace if worn.
(b) Check that the driven pulley is tight on its shaft.
(c) Tighten all bolts on grizzly bars and liners.
(d) Check and if necessary adjust the feed end spring/snubber assembly, where fitted.
(e) Check the vee belt tension.

6 monthly: (or more frequently if thought desirable)
(a) Drain, flush and refill with new oil.
(b) Clean oil sight glasses and replace if discoloured.
(c) Check and cover oil seals. Replace if leaking.
(d) Check setscrews retaining end covers. Tighten if necessary.
(e) Check bolts retaining vibrating unit to side plates. Tighten if necessary.

(f) Remove centre cover of vibrating unit and check setscrews retaining counterweights. Tighten and rewire if necessary. Replace gasket of centre cover if damaged.

(g) Clean all breather valves. Replace if damaged.

(h) Replace any weakened support springs.

Suggested site stock spare parts for an apron feeder:

(a) One of each size of bearing or plummer block.

(b) Selection of repair links for the drive chain or a complete chain.

(c) One set of vee belts where fitted.

Suggested site stock spare parts for a vibrating grizzly feeder:

(a) Pan liners and bolts.

(b) Frame liners and bolts.

(c) At least one grizzly bar and a full set of bolts for all grizzly bars.

(d) Selection of springs.

(e) One set of vee belts.

(f) One set of gaskets, "O" rings and oil seals.

(g) One set of oil level indicators and breathers.

(h) Friction blocks where fitted.

(2) *Electric vibrating feeders*

These feeders are occasionally encountered on comparatively small plants and are also popular for use over reclamation conveyors in surge-pile tunnels.

Many types are available but most use the same operating principle in which electrical impulses are passed through a stator and coils forming a magnetic circuit to create a sequence of interrupted magnetic pulls on an armature which is linked to the pan. The amplitude of vibration is usually infinitely variable within pre-set limits.

Very little attention is normally needed but whenever repair work is carried out make quite sure that the electrical supply is disconnected.

There is an air gap between the armature and the stator facings; do not allow dust etc. to accumulate in this gap, particularly if there is water about.

Suggested site stock spare parts for electric vibrating feeders
One coil set.
One ammeter.
One variable resistance.
One torovolt transformer.
One rectifier unit.
One set input fuses.
One set output fuses.
One set variable resistance brushes.
One set torovolt brushes.

(3) *Reciprocating plate feeders*

These are sometimes used on the smaller plants and comprise basically a plate moved back and forth by one or more eccentrics, variable for stroke and hence feed rate. They are generally suitable for comparatively light duties only, are simple in construction and are not particularly demanding on maintenance.

Plain, babbitted (white metal) bearings are often used for the shafts and eccentrics.

Routine operating and maintenance points

(a) Always disconnect the main power source to an electrically driven feeder before starting any maintenance work.

(b) Do not attempt adjustments with the feeder running.

(c) Do not operate the feeder without the drive guards in position.

(d) Give each grease nipple two or three shots of grease at four-hourly intervals.

(e) The grease cups, particularly on the eccentrics, should be given two or three turns every four hours − or more frequently if there are signs of overheating − and refilled as necessary.

(f) If there is jockey pinion adjustment, maintain the chain at a tension just short of tight.

(g) The chain is usually lubricated by a drip-feed lubricator which must be kept clean, undamaged and charged with suitable oil.

(h) Replace the wearing plates in good time. If these plates become distorted they will restrict the flow of stone and there will be wear on the main structure of the feeder.

(i) At monthly intervals remove the chain drive case and clean out any accumulation of dust and old oil. Take the opportunity of cleaning the chain with brush and paraffin or similar.

(j) If oversize or slabby rocks "hang up" in the feeder do not attempt to clear them by hand – use a bar. If a rock is too large to enter the crusher stop the plant by means of the engine clutch before lifting the rock out of the feeder hopper.

(k) Do not allow spilled rock to accumulate on the feeder platform.

Suggested site stock spare parts for a reciprocating plate feeder:

(a) Selection of repair links for the drive chain or a complete chain.

(b) A drip-feed chain lubricator, if fitted.

(c) A bush for the jockey pinion, if fitted.

(d) A set of vee belts if used in addition to a drive chain.

(e) A set of liners and bolts.

(4) *Jaw crushers*

Single toggle jaw crushers are now more commonly encountered than the double toggle version. With either type it is essential for maximum output that the feed should be small enough to flow from the feeder into the crusher mouth without bridging, as discussed earlier.

It is also very desirable that all material smaller than the crusher discharge setting is removed from the feed and this is the function of the grizzly, either stationary or as part of the vibrating feeder. These "fines" tend to cause a packing action within the crushing chamber, resulting in heavy stresses in the machine, accelerated wear on the bearings and toggle plate and reduced life of the manganese steel jaw plates.

The feed of rock should be regulated so that the crushing chamber is always full. Wet, sticky material will reduce output.

The following routine operating and maintenance points refer generally to single toggle roller bearing jaw crushers but can with discretion be related to double toggle jaw crushers.

66

1. SPRING CAPS   5. TOGGLE PLATE
2. SPRINGS   6. SPRING BRACKETS
3. TENSION RODS   7. TOGGLE BEAM
4. JAWSTOCK

Fig. 21   THE SPRING AND TENSION ROD ARRANGEMENT OF A
TYPICAL SINGLE TOGGLE JAW CRUSHER

Photo courtesy Pegson Ltd

For fuller and more specific instructions you must of course refer to the manufacturer's manual.

(a) Always disconnect the main power source to an electrically driven crusher before starting any maintenance work.

(b) Do not operate the crusher unless the guards are in position on flywheels and drives.

(c) Where an electrically driven feeder is incorporated in the plant, disconnect the power source before working on the crusher.

(d) *Station a man on the ramp to ensure that no dump truck tips during maintenance work on the crusher.*

(e) If an oversize rock reaches the crusher mouth either remove it with a grapple (if available) or split it with hammer and wedge – *not* with explosives. Do *not* attempt this whilst either the crusher or feeder is running. Take the precaution in (d) above.

(f) When the machine is running empty there should be virtually no mechanical noise. If such noises develop, stop the crusher and investigate.

(g) Adjust the spring(s) so that the crusher runs without toggle plate knock. This is important; if the spring is insufficiently compressed the toggle plate will bump and may drop out. If too tight the spring or the tension rod may break.

(h) Remember that the jaw plates are often reversible top

67

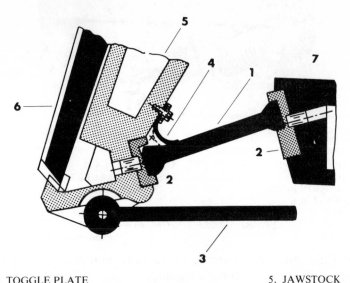

1. TOGGLE PLATE
2. RENEWABLE SEATS WITH RETAINING PINS
3. TENSION ROD
4. CANVAS PROTECTING APRON
5. JAWSTOCK
6. SWING JAW
7. ADJUSTMENT

Fig. 22   TOGGLE ARRANGEMENT OF A SINGLE TOGGLE JAW
CRUSHER

for bottom; reverse each casting before the corrugations are completely worn away from the bottom section of the face. Never allow the jaws to wear to a concave profile; this will cause the rock to "hang up" in the crushing chamber thus (a) reducing capacity (b) increasing power consumption (c) causing undesirable stresses in the bearings and toggle plates.

(i) Bear in mind that the toggle plate, as well as transmitting the motion of the crusher, usually also functions as a safety device to protect other parts of the machine against overload. If a toggle plate breaks the cause is usually to be found in one of the overload conditions described in these notes. *Investigate at once.*

(j) Never allow uncrushable material to enter the crusher. This includes timber as well as digger teeth, drill heads, hammer heads etc.

(k) Keep all bolts tight, particularly those retaining the jaws plates. If jaws become loose, damage may be

1. TOGGLE BEAM CLAMPING BOLTS   3. TOGGLE BEAM
2. SHIMS   4. TURNBUCKLE

Fig. 23   SCREW TURNBUCKLE ADJUSTMENT OF A SINGLE TOGGLE
JAW CRUSHER

Photo courtesy Pegson Ltd

caused to machined seating pads on the jawstock and main frame.

(l)   Sideways movement of the swing jaw is often prevented by keys. If these keys become detached or worn, fit new ones without delay.

(m)   Always use factory-made adjusting shims and be sure to fit the same total thickness each side. Make-shift, uneven shims can cause cross-loading of the toggle plate and subsequent failure.

(n)   On oil-lubricated machines replace immediately any flexible pipes which become chafed or collapsed, using only oil-resistant hose. Do not allow the oil alarm system to fall into disuse.

(o)   On grease lubricated machines, do not overgrease. Generally one or two shots of grease monthly per bearing will suffice, but refer to the operating and maintenance manual for your particular machine.

(p)   Excessive grease will cause the bearing to overheat and the bearing will eject grease until the correct quantity remains in the cavities, after which the temperature will drop. If a bearing is overheating and *not* ejecting grease stop the machine and investigate. This condition will often be accompanied by a clicking or grinding noise from the bearing.

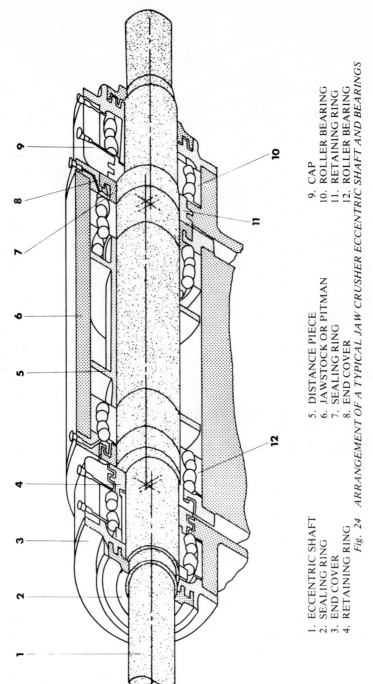

1. ECCENTRIC SHAFT
2. SEALING RING
3. END COVER
4. RETAINING RING
5. DISTANCE PIECE
6. JAWSTOCK OR PITMAN
7. SEALING RING
8. END COVER
9. CAP
10. ROLLER BEARING
11. RETAINING RING
12. ROLLER BEARING

Fig. 24   ARRANGEMENT OF A TYPICAL JAW CRUSHER ECCENTRIC SHAFT AND BEARINGS

70

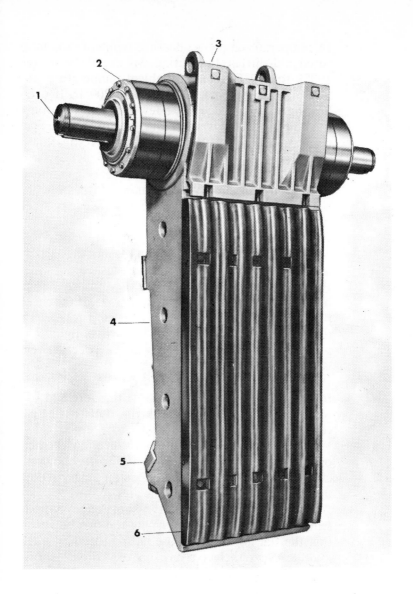

1. ECCENTRIC SHAFT   4. JAWSTOCK
2. OUTER BEARING ASSEMBLY   5. TOGGLE SEAT
3. RENEWABLE WEAR PLATE   6. SWING JAW

Fig. 25 JAWSTOCK ASSEMBLY OF A SINGLE TOGGLE ROLLER BEARING JAW CRUSHER

Photo courtesy Pegson Ltd

71

(q) In temperate climates, bearing temperature, measured mid-shift at the housing, will usually be around 100°F to 130°F. In high ambient temperatures the housing may reach 180°F, peaking 200°F. If overheating is suspected, spray a few drops of water onto the housing; if the water steams on contact the temperature is above 212°F and further investigation is indicated.

(r) Never, however, attempt to reduce bearing temperature, even temporarily, by running water onto the housing. The sudden temperature change may crack or crystallise some part of the assembly.

(s) When the crusher is operating, be alert to the possibility of rocks being ejected from the mouth, particularly if hard, round stone is being crushed.

(t) Pay particular attention to keeping the cheek plate bolts tight and replace any missing bolts. If the cheek plates become loose, dust and grit will pack behind them making it impossible to keep them tight. The plates may eventually rub the swing jaw, become very hot and break.

(u) Prolonged impacting of stone directly on the top front of the jawstock may distort the bore and will wear the casting. If a jawstock guard plate is fitted, replace or repair it when worn.

(v) When carrying out welding repairs on a jaw crusher, or indeed on any machine, be sure to fix the earth clamp in such a position that the welding current does not pass through the roller bearings.

(w) A broken drawback spring can sometimes continue temporarily in service by turning the two halves so that the flat ground ends are butted together in the middle. Nevertheless a broken or "tired" spring should be replaced as soon as possible or damage to the tension rod may result.

(x) If the toggle plate moves sideways replace at once the broken or worn retaining pin or block and recentre the plate. Do not allow the plate to rub the crusher side wall which may break the plate and will increase power consumption. Take the opportunity of identifying and correcting the cause of the toggle plate moving sideways and overcoming the retainer. This will usually be (a) heavy wear on the toggle plate and/or toggle seats, or (b) misaligned or loose adjustment mechanism.

Suggested site stock spare parts for a jaw crusher:
(a) Fixed jaw plate and fixings
(b) Swing jaw plate and fixings
(c) Side cheek plates and fixings
    Realistic stocks of the above parts will depend on the local rate of usage.
(d) Toggle plate
(e) Toggle seats
(f) Oil or grease seals
(g) Spring(s)
(h) Pins and bushes for the tension rod(s)
(i) Vee belts − a matched set.

As with feeders a preventive maintenance routine is to be recommended and the following typical schedule should be considered:

Daily:
(a) Check that vee belts are correctly located in grooves of motor pulley and flywheel.
(b) Check that guards are in position and secure.
(c) Check for toggle plate knock. Adjust the spring if necessary.
Weekly:
(a) On grease-lubricated crushers inject approximately 1 oz. of grease via each nipple.
(b) On oil lubricated machines check the oil level.
(c) Check fixed jaw bolts. Tighten if necessary.
    Replace any missing bolts.
(d) Check swing jaw bolts. Tighten if necessary.
    Replace any missing bolts.
(e) Check side cheek bolts. Tighten if necessary.
    Replace any missing bolts.
(f) Check the adjusting mechanism for tightness.
(g) Check that toggle plate is intact and located centrally in seats. If there is sideways displacement, investigate and rectify.

Monthly:
(a) Check nuts retaining main frame bearing caps. Tighten if necessary.
(b) Check studs retaining shaft end caps, tighten if necessary.
(c) Check vee belt tension. Adjust if necessary.
(d) Check bolts retaining jaw guard (where fitted). Tighten if necessary.
(e) Check bushes in tension rods and sleeves.
   Replace if excessively worn.

(5) *Cone crushers*
   Cone crushers for secondary and tertiary reduction duties are available in a wide variety of types and sizes. Operating principles vary as do the types of bearings incorporated. Setting, adjustment and overload protection functions can be mechanical or hydraulic.
   For maximum output and trouble-free service it is essential to familiarise yourself with your cone crushers, particularly by studying the service manuals and as with jaw crushers the following notes on routine operation and maintenance are guidelines only:
(a) Keep the cone crusher well fed. There must always be sufficient stone in the crushing chamber to prevent the head from spinning. The crushing members, i.e. the head and bowl liners, are manganese steel castings which depend for life-effectiveness on work-hardening by impact. Abrasion is bad for manganese steel and if the head is allowed to spin, abrasion is inevitable. On most machines the head will "creep" in anti-rotation during crushing; this is normal and acceptable.
   All cone crushers are designed to be choke-fed and will discharge a better product shape when this condition is achieved and maintained.
(b) Modern cone crushers are almost invariably oil-lubricated and a lubrication failure can be very expensive in terms of down-time and bearing replacements. Study the manufacture's instructions and follow them closely. Pay particular attention to the pressure gauge, the filters and to oil-change periods. If you have any doubts at any time, err on the safe side.
   There will probably be an alarm system to guard

against (1) excessively high oil temperature and (2) low oil pressure.

Do not allow this equipment to fall into disuse through, for example, blown fuses, damaged wires or pipes, dust-clogged syren. Generally the oil should not be more than hand-hot; if this condition is not maintained or if the pressure falls below the specified limit stop the crusher and investigate, whether or not the alarm system has reacted.

The oil tank will usually be provided with an oil level gauge. Check this gauge at least once every shift and top up if the oil level does not appear in the gauge. Checking the gauge is easier if the glass is kept clean; if it is a tube type gauge you can ensure this by fitting over it a finger cut from an old industrial glove.

(c) Never be tempted to run a cone crusher until the head and/or bowl liners are worn right through. This can cause (i) damage to the main castings and (ii) damage to the bearings by pieces of manganese steel becoming detached and nipped.

(d) To measure the discharge setting attach a ball of lead foil or firm clay, larger in diameter than the known "closed" side setting, to the end of a length of wire or strong cord. Lower the ball into the crusher chamber on the "open" side until it is just above the bottom edge of the head liner. Rotate the flywheel by hand until the head closes and reopens on the ball. The compressed thickness wiil be the setting of the discharge opening. Alternatively, with the crusher running and a light feed of stone drop a loose ball of lead into the chamber to pass through the machine and retrieve it from the discharge conveyor. Take care when picking up the flattened lump of lead − it will be hot. Do not attempt this second method with no stone and the head spinning; the result will be misleading.

(e) On some cone crushers it is possible to extend the life of the drive gears by reversing the rotation of the flywheel. Before attempting this, consult the manufacturer's instructions; other adjustments may be needed.

(f) Check periodically the compressed length of the springs with which many cone crushers are fitted. The required length will be specified in the service manual and should be the same on each spring. Their function is usually to protect the machine against uncrushable material and excessive "packing". Slight "breathing" may be acceptable but if there is pronounced bouncing first verify the spring compression. If the compression is correct and the bouncing persists investigate immediately — the cause will usually be found in unsatisfactory feed conditions.

Backing cone crusher liners

Generally the bowl liner, which is a manganese steel casting, will be secured by wedges or bolts. The head liner, also manganese steel, will have a locking device of, for example, a large steel nut and cap. There will be a cavity between the back of the liner and the main casting and the cavity must be filled by a backing to support the manganese steel and to provide a certain amount of resilience to accommodate distortion during crushing.

The traditional backing material was zinc, poured in molten form, but this was a difficult and somewhat dangerous operation. The current trend is to use a two-part plastic material plus a release agent to prevent adhesion where it is not required. There are several brand names for this type of backing material and it is important to ensure that the type you use is guaranteed for use as "crusher backing".

Preventive maintenance requirements for cone crushers will vary widely according to the manufacturer's instructions but the following points are worth consideration:

Daily:
(a) Check that the vee belts are correctly located in the grooves of the drive and driven pulleys.
(b) Check that the drive guard is in position and secure.
(c) Check that the oil alarm system is functioning.
(d) Check that the oil level appears in the gauge.
(e) Check for and rectify any excessive spring movement.
(f) Check that the oil pressure is correct as specified.

Weekly:
(a) Attend to any leaking oil pipes and joints.

Monthly:
(a) Clean the oil filters. Check for the presence of metal particles and investigate if any are found.
(b) Check vee belt tension.

6 Monthly: (or more frequently if thought desirable)
(a) Drain, flush and refill the oil tank.

Suggested site stock spare parts for a cone crusher:
(a) Manganese steel liners with fixings and backing material The quantities will be determined by your local usage.
(b) Seals, piston rings, bellows and gaskets as fitted in your machine.
(c) Oil filters.
(d) Oil pressure gauge.
(e) Oil level gauge.
(f) Vee belts − a matched set.

(6) *Vibrating screens*
Of all the main machines incorporated in a processing plant the vibrating screen varies most in type and detail. The majority however are mechanically vibrated and there are two main groups:

(a) Horizontal, usually vibrated by a twin-shaft mechanism.
(b) Inclined, usually with one shaft and one or more eccentric masses. These counterweights are almost always variable and may have an automatic adjustment mechanism.

The following routine operating and maintenance points are widely applicable:
(a) Always disconnect the main power source to the electric motor before starting any maintenance work.
(b) Stop the feed conveyor before working on a screen.
(c) Do not operate the screen unless the drive guard, balance wheel guard and shaft extension guard (if used) are in position and secure.
(d) If there are oil level sight indicators check them at the start of each shift and top up if necessary. If the indicators become discoloured, fit new ones, wash the old ones with a solvent and put them into store.

(e) This is a vibrating machine so replace without delay any missing bolts or screws.
(f) Replace any worm feed box or discharge lip liners without delay. Old conveyor belting fitted into the feed box will cushion the fall of stone and reduce wear.
(g) Replace without delay any broken or "tired" support springs.
(h) Each nipple on a grease lubricated screen will probably need a shot of grease per day or as instructed by the manufacturer. Over-greasing is to be avoided however since the churning effect may cause starting problems, particularly in cold climates.
(i) . If the counterweights are automatic a broken retractor spring or rubber strand will produce a distinctive knock from the balance wheel, will cause rough running during starting and stopping and must be replaced without delay.
(j) Mesh cloths must be tensioned to eliminate any whip or drumming and must bear firmly on their supports which are usually rubber-capped. Any movement within the mesh cloth or of the mesh cloth relative to its fixings and supports will result in early failure. When a mesh does eventually fail, the cause can often be ascertained from the condition of the broken wire ends. If the wire diameter is not significantly reduced, failure probably resulted from insufficient tension. Thinned wire has probably failed from normal wear. Screening surfaces are considered in more detail later in this section.

Here is a suggested preventive maintenance schedule for vibrating screens:

Daily:
(a) Check that the vee belts are correctly located in the grooves of both pulleys.
(b) Check that all guards are in position and secure.
(c) Watch for loose or missing mesh tensioning fixings. Tighten or replace as necessary.
(d) Check oil levels or attend to greasing as required by the manufacturer.

Weekly:
(a) Tighten all mesh tensioning fixings.

(b) Replace any worn or damaged rubber cappings beneath the meshes.

(c) Tighten all bolts on liners and discharge lips.

(d) Check vee belt tension.

6 Monthly: (or more frequently if thought desirable)

(a) Where appropriate remove the balance wheel covers and inspect the bushes and springs of automatic counterweights.

(b) On oil lubricated vibrating mechanisms drain, flush and fill with new oil.

(c) Check all accessible vibrating mechanism bolts for tightness.

(d) Clean any breather valves and replace any damaged ones.

(e) Clean any oil level sight gauges.

(f) Replace any weakened support springs.

(7) *Screen deck media, tensioning and fitting*

Probably the most important part of any vibrating screen and often the most neglected is the screening media of which the main types are:

(a) Woven wire mesh

(b) Welded wire mesh

(c) Perforated plate

(d) Moulded polymer or rubber

On the so-called flat vibrating screens the sizing media will consist of one, two, three or occasionally four decks. The decks on a particular screen will usually be of the same overall dimensions, depending on the rated capacity and hence on the nominal size of the machine, but for ease of fitting, maintenance and replacement the panels of mesh etc. will be of a shorter length and possibly width. In practice the most convenient length is around 1.5 metres and the width will normally be the nominal width of the screen deck since this arrangement simplifies securing and tensioning the panel. The screening media is of course intimately in contact with the stone, gravel, sand or whatever is being graded and must not only sort the particle sizes but must also transport the product from the feed end to the discharge end of the deck. To achieve both requirements involves rapid displacement of the particles, both horizontally and vertically, and the deck is consequently subjected to both abrasion and impact by what is usually a very hard, angular material. The most common construction is wire mesh of either woven or welded construction, with the diameters of the individual wires related to the size of the apertures between them. On woven meshes the wires in both directions, the warp and the weft, are crimped, that is to say, bent over each other. The application of tension is very important to most types of screening media; it is particularly important on woven wire since there must be no possibility of the crossing wires becoming slack and rubbing against each other. Similarly the wires must not be allowed to rub or vibrate against whatever structure is employed to support them across and along the deck. It is for this reason that the support members are usually capped with rubber mouldings. Welded wire construction is often used, particularly for apertures of 30 mm and above. The wires are welded at each intersection; tension is again important but more difficult to apply and where welded meshes are used attention to the rubber capping is particularly important.

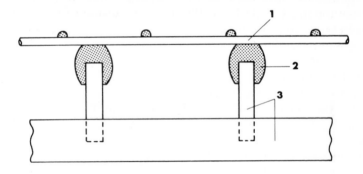

1. WIRE MESH   3. STEEL SUPPORT FRAME
2. RUBBER CAPPING
Fig. 26   TYPICAL SUPPORT FOR WIRE MESH PANEL

It must be emphasised again that whatever type of wire mesh is used, tensioning is absolutely vital. Each panel must be drum-tight.

The wires from which the screen mats are made can be formed from a variety of metals. Where economy and strength are the main factors, mild steel will meet normal requirements. To prevent rust the wire can be galvanised or otherwise plated but, of course, on quarry duties the plating never lasts long.

Where resistance to corrosion is needed, stainless steel wire can be used, obviously at increased cost, and provided the correct grade is chosen, the wire is very tough.

Phosphor bronze is sometimes used for the same reasons for weaving mats with very small apertures. When the wire is electrically heated, as in some special duties, copper is often favoured.

Monel metal is also suitable for electric heating whilst nickel is sometimes used where there is a possibility of chemical attack.

The apertures formed by the crossing wires are generally square and of course the size of the apertures, bearing in mind the angle at which the screen is mounted, determines the size of particle which can pass through. Sometimes a "slot" shaped aperture is used, particularly where "pegging" or "blinding" of the deck is encountered.

This "pegging" or "blinding" is fairly common. Blinding results from moist fine material, that is to say, sand and clay, building up around the mesh openings. Pegging is due to splinters or elongated pieces of stone standing on end in the mesh openings. Blinding usually starts around the pegged pieces of stone and if pegging can be prevented the mesh can often be kept clear of blinding.

One method of preventing excessive pegging is by using a ball deck. Such an arrangement consists of a second, coarser

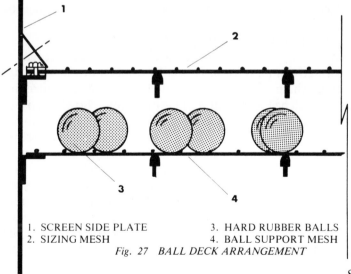

1. SCREEN SIDE PLATE   3. HARD RUBBER BALLS
2. SIZING MESH   4. BALL SUPPORT MESH

Fig. 27 BALL DECK ARRANGEMENT

retaining mesh a few inches below the sizing deck. Hard rubber balls placed between these two meshes are caused to bounce by the vibration and knock the sizing mesh continuously from below. This will only succeed if the upper mesh is light enough to be displaced by the bumping of the balls and invariably increases the rate of wear on the upper mesh.

Where sand alone, or perhaps with a small proportion of gravel, is being screened, what are known as individual wire or piano wire decks are used. The deck looks very like the strung frame of a piano and the distance between the wires determines the particle size. Each wire is individually tensioned along the length, not the width of the screen. By virtue of the fact that the cross wires are dispensed with and also that the wires are free to vibrate at their own individual vibration frequency, blinding and pegging can be largely eliminated.

Screen mats are now available in a variety of moulded polymers. These mats have a much greater resistance to abrasion and impact compared to metal panels. They are less prone to pegging and blinding and they will not corrode in the wet conditions frequently encountered in the grading of aggregates. They reduce the degradation of friable materials and their noise level is substantially lower than that of metal surfaces. The mats can be rolled, carried and manoeuvred comparatively easily into restricted spaces.

Initial cost is higher than for metal panels but longer life and reduced down-time are claimed to cut screening costs per ton.

Edge treatment of screening media

Whatever the type of material used for the construction of the perforated panel, tension is usually applied at the side edges although the panel may also be clamped along its centre-line. The aim will be to generate a single or double "bow" across the width and this curve greatly facilitates the application of tension. Tension edge treatments vary although most rely on bolts passing through the screen side-plates, either at an angle or normal to the plate surface. Angled bolts will pass through slotted holes in the side plate and will be provided with taper or rocker washers on the outside. There are usually double nuts or self-locking nuts.

The illustrations which follow show some common edge treatments and tensioning arrangements.

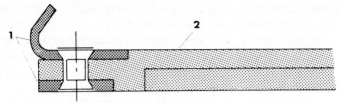

1. STEEL 2. POLYURETHANE

Fig. 28 POLYURETHANE MAT EDGE TREATMENT

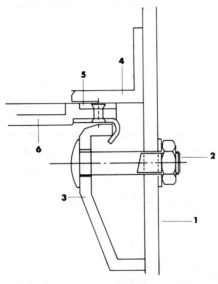

1. MACHINE SIDE PLATE   4. LEDGE ANGLE
2. TENSION BOLT   5. EDGE PREPARATION
3. TENSION PLATE – TYPE 2   6. POLYURETHANE SCREEN MAT

Fig. 29 POLYURETHANE MAT TENSION SYSTEM

MANGANESE STEEL WEARING PARTS

Crusher jaws, cheek plates, cone liners and head liners are required to be cast in manganese steel.

The appropriate British Standard Specification is BS 3100: 1976 which provides for the following chemical composition:

Element	% min.	% max.
Carbon	1.00	1.25*
Silicon	–	1.00
Manganese	11.00	–
Phosphorus	–	0.070
Sulphur	–	0.060

*For special applications, by agreement between the manufacturer and the purchaser, the maximum carbon content may be increased to 1.35%.

Heat treatment is a very important requirement which can hardly be overstated. BS3100 demands:

All castings shall be supplied in the water quenched condition. Water quenching shall be carried out from a temperature of not less than 1000°C.

BS3100 also states that:

The castings shall be free from adhering sand and scale and free from defects capable of detection by the method of inspection (i.e. visual) and of a severity level greater than the acceptance standards agreed between the manufacturer and the purchaser.

The standard further states that:

Castings exhibiting unacceptable defects may be rectified by *(welding). Grinding or other means may be used to remove superficial surface defects.

*The clause on welding states:

Unless otherwise specified by the purchaser on the enquiry and order, castings may be rectified by welding without the previous sanction of the purchaser. Rectification by welding shall be carried out in accordance with BS4570: Part 1.

Work hardening

This is an important property of correctly processed manganese steel. As cast the metal is comparatively soft; the austenitic structure is generated by the heat treatment and it is this structure which gives the steel its capability of work-hardening. The initial hardness of the surface layer is around 220 Brinnell; with impact the austenitic surface layer is transformed to martensite of approximately 550 Brinnell. As this hardface layer is worn away in service, the freshly exposed surface continually becomes hardened by impact and the casting thus combines the advantages of the toughness of the austenitic core with the hardness of the martensitic surface.

If the casting has been correctly heat-treated there is inevitably a work-hardening factor in the subsequent machining. This calls for slow machining feeds (i.e. speeds) and deep cuts on a very rigid machine tool, e.g. borer, lathe or planer.

Dimensional requirements

Ideally there should be a perfect interface between the manganese steel casting and its supporting member i.e.

A swing jaw to the jawstock.

A fixed jaw to the crusher body.

84

A head liner to the crusher head.
A bowl liner to the crusher support bowl.

Any relative movement during crushing will inevitably abrade and deform the supporting member leading to:
(a) Possible cracking of the manganese steel due to insufficient support.
(b) Loose manganese steel with a worsening of the deformation.
(c) Costly repairs to the supporting member.

Crusher jaws should have no bowing and at least 80% flatness on the rear face with no high or low areas exceeding 2 mm.

Cone crusher liners must have absolutely accurate tapers on the contact bands. Face and back surfaces, machined or unmachined, should be concentric about the centre-line with no appreciable differences in thickness:

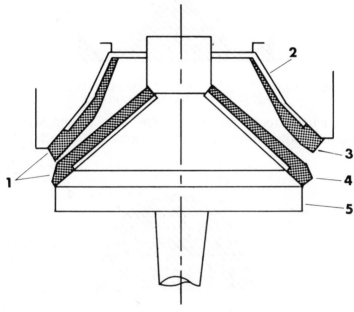

1. INTERFACES   4. HEAD LINER
2. SUPPORT BOWL   5. CONE HEAD
3. BOWL LINER

Fig. 30   JAW CRUSHER TOOTH PROFILES

Jaw crusher tooth profiles
Ideally the line of impact of the rock on the jaw surface should be at a right-angle for best work-hardening. This is particularly important on primary jaw crushers.

This ideal condition could only be achieved with flat jaws but at a penalty on efficiency and throughput since there would be no "splitting" action on the stone.

A pointed tooth profile provides the best "splitting" action but:

(a) The angle of impact results in a skidding effect i.e abrasion at the expense of impact.

(b) The line contact at the tip or peak of the tooth, depending on the nature of the stone, more or less quickly degenerates into an irregular curved profile, with gradually increasing impact efficiency.

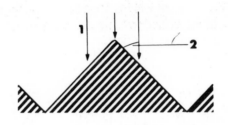

1. LINE OF IMPACT   2. ANGLE OF IMPACT ON FLANK
Fig. 31   A POINTED TOOTH PROFILE

A curved tooth profile combines the advantages of a splitting action with a better impact angle on the rock.

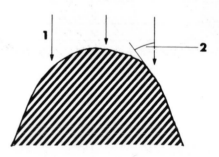

1. LINE OF IMPACT   2. ANGLE OF IMPACT ON FLANK
Fig. 32   A CURVED TOOTH PROFILE

The valleys between the teeth are as important to efficient crushing as the peaks. When a lump of rock is crushed it expands to occupy more space. The fragments must be able to descend quickly within the crushing chamber either to the next "nipping" stage or to pass out of the chamber as sized rock.

The valleys must be shaped to facilitate this movement and a curved surface, with no angles, provides the best channel. Thus although a pointed tooth profile will wear to a curve, presenting a reasonably good face to the rock, the angled valley remains to hinder transit of the particles.

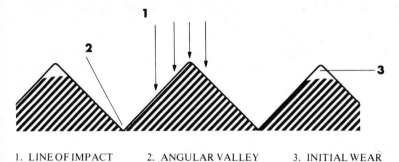

1. LINE OF IMPACT   2. ANGULAR VALLEY   3. INITIAL WEAR
Fig. 33

For the great majority of crushing duties therefore a double curve or sine tooth profile is to be preferred.

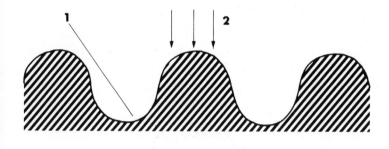

1. CURVED VALLEY   2. LINE OF IMPACT
Fig. 34   A SINE TOOTH PROFILE

THE DUST HAZARD

(The co-operation of DEMISTER AB, Malmo, Sweden, is acknowledged in the preparation of this section.)

By the very nature of the operations involved and the raw materials processed, quarries are prolific sources of dust. "Dust" is a term used for any particulate material having a particle size smaller than 100 micron (100 micrometre = ca. 140 mesh). Mineral particles above 100 micron do not normally remain airborne for long periods. Dust below 5 micron is respirable and constitutes the main health hazard causing lung diseases such as pneumoconiosis.

As an insight into this hazardous micro-world of dust consider that:

(a) The lower limit of unaided visibility is 45 micron.
(b) The diameter of a human hair is 50 micron.
(c) To be visible to the naked eye the dangerous 5 micron particle would have to be increased in its mass by around 800 times.
(d) When silica is involved, environmental health authorities commonly specify a permissible dust content in the working environment of one milligramme or less per cubic metre of dry air. If all of the particles in the 1 mg. sample were of 5 micron size, the sample would contain 6 million particles. In reality, however, the particle size distribution would be such that the number of particles in the sample would be between 1.2 and 1.4 milliard. Nevertheless the whole sample could remain unnoticed under a finger nail and each particle is potentially dangerous if inhaled.

Typical statutory regulations relating to quarries demand:

(a) Where . . . there is given off dust . . . likely to be injurious to persons employed, it shall be the duty of every manager . . . to ensure that there are steps taken . . . necessary to protect those persons against inhalation of the dust.
(b) that the entry of dust into the air or its accumulation in any place . . . is minimised by . . . steps . . . taken as near as possible to the point of origin of the dust.
(c) that any dust which enters the air is trapped . . .
(d) that any dust . . . not prevented from accumulating is systematically cleared up and removed.

In the average small quarrying operation dust will be generated and become airborne at several locations. Although total elimination of respirable dust may be considered impracticable, particularly on a mobile crushing and screening plant, simple economics (if no other consideration) demand that the workmen be afforded some protection.

Dust masks are readily available but are not generally considered as complete protection and in hot climates can be uncomfortable to wear. Nevertheless, they ought to be provided and the workmen instructed in their use, management and supervisory staff setting an example.

Other measures may have to be taken into account, for example, federal or local regulations, capital availability, mechanical practicability etc. What follows represents a pragmatic approach to conditions which will vary widely from quarry to quarry:

(1) First and most obviously, consider the prevailing wind. Site workshops, mess buildings, offices, generator houses etc. should be as far as practicable up-wind of the processing plant.

(2) Roadways from the works entrance to the loading points and any other areas where there is a regular movement of vehicles should, where practicable, be hard-surfaced and kept clean. In dry climates, regular watering of the surface will lay the dust; drainage will need attention but will not normally be difficult.

(3) The drilling rig is a potential dust source, unless wet flushing is employed. If air flushing is used, the prevailing wind may not always protect the crew and masks should be provided.

(4) The dumping of rock into the primary crusher hopper often impels into the atmosphere large amounts of dust, some of which, in the range 100 micron plus, will quite quickly precipitate. The prevailing wind may well aggravate the situation by driving the dust into the 180° arc around the plant, exactly where men are likely to be working and wherever the wind, the man on the primary crusher platform will momentarily at any rate, be enveloped in dust. This man should always have a mask available and must know how to use it. In practical terms the only palliative available at the dump hopper is suppression by water where the dust is subjected to a fine water spray at a stage

when it is still adhering to the surfaces of the rock. It is thus kept down before it has a chance of becoming airborne.

This latter point is important but it must be understood that it is literally impossible at a dump hopper to prevent some of the dust rising and dust particles below 5 micron will not be caught by any liquid spray, even when a wetting additive is used. The essential condition for the capture of the hazardous 5 micron and smaller particles is that the airborne dust particle must collide with an airborne water particle within the very short contact time available. A brief technical digression will explain why this condition is not achieved to any significant extent.

For the collision to occur, the difference between the mass of the particle and the mass of the water droplet must not exceed 400 to 500 otherwise the smaller body evades the larger. This sketch shows why:

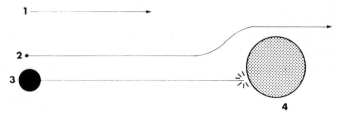

1. AIR STREAM
2. DIRECTION OF 5 MICRON PARTICLES
3. 45 MICRON PARTICLE COLLIDES
4. 110 MICRON WATER DROPLET

Fig. 35

Here we see two particles carried in an air current, one of 5 micron size, the other of 45 micron size. Only the larger particle has a mass big enough not to follow the air current past the 110 micron water droplet but to maintain its course and collide. The small particle is so light that it follows the air current and no collision occurs.

The finest water spray which can be economically produced has a droplet size from 35 to 150 micron. Even the smallest droplet of this aerosol has a mass 350 times larger than the 5 micron particle which we wish to suppress. When it is considered that at least 95% of the solid particles are smaller than 5 micron and at least 95% of

90

the droplets are larger than 35 micron it can be seen that the probability of a high frequency of collisions is poor. Furthermore a particle and a droplet, in order to collide, must actually be on a collision course and this depends on the relative densities of the particles and droplets within the collision area. Of our own sample of 1 mg we have, say, 500 million particles, so each particle has its own free space of 2 cubic mm. Thus the average distance from one particle to the next is around 850 times its own diameter.

It is therefore quite impossible to suppress all dust at the dump hopper and masks should be provided for all personnel at work in that area.

(5) There will always be a certain amount of dust emission from the feeder and from the primary crusher. Wetting the rock at the dump hopper will help and it may be useful to fit a further spray bar over the crusher mouth. Even so, any men employed around the primary plant should use masks.

(6) On-plant, inter-plant and stockpile conveyors will emit dust, particularly at transfer-points and from belt runs exposed to a cross wind. Water sprays can be used but unless the amount of water is kept to a minimum, in which

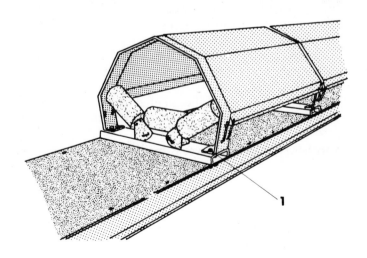

1. HOOP SECURED BY TRANSOME FIXING BOLTS
Fig. 36   COMBINED SHEET METAL COVER AND WINDBOARDS

case efficiency is doubtful, there will be sludge problems around the tail drums. Windboards are a useful asset and can be combined with sheet metal covers in severe conditions:

As a further refinement a steel box cover can be used to encapsulate the entire conveyor:

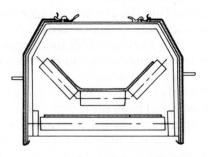

Fig. 37   A STEEL BOX COVER

Drawbacks to box covers are:

(a) Moving machine parts remain inside the cover and are thus exposed to and work in a dusty atmosphere, resulting in increased wear.

(b) Lubrication of moving parts is difficult. Points to be frequently lubricated would have to be connected by pipes to nipples outside the cover.

(c) Spillage and settled dust can cause problems and have to be cleared out.

(d) Steel panels may be dented and distorted when handled.

Another type of box cover is made of fabric:

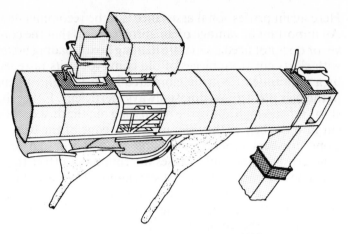

Fig. 38   A COVERED CONVEYOR

The design and manufacture of box covers is a specialist business and is not normally undertaken without consultation with a professional engineer.

The feed boot of a conveyor can be sealed by an apron cover:

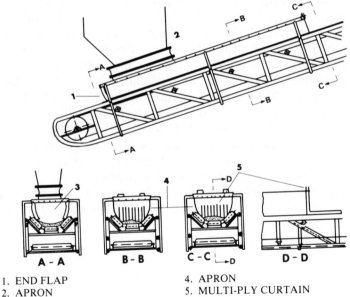

1. END FLAP
2. APRON
3. END FLAP

4. APRON
5. MULTI-PLY CURTAIN

Fig. 39   FEED BOOT COVERS

Here again professional assistance is to be recommended. An important advantage of an apron cover is that the conveyor does not need a separate loading box. Loading boxes with skirts cut from pieces of old conveyor belts fastened by a multitude of bolts in elongated holes for adjustment to the belt surface are almost always inefficient. Either the skirts remain high above the belt letting out the dust or they cut into the belt or they slip under the edge of the belt.

(7) Cone crushers generate dust and since the design of a cone crusher inherently causes a "chimney" effect these machines are often the worst offenders. They are, however, comparatively easy to deal with by:

   (a) Sealing the discharge chute as far as practicable to minimise the up-draft.

   (b) Sealing the feed chute and the feed ring. A light steel grid can be shaped to the top of the feed ring (wide aperture screen mesh can be used) and will support rubber sheet held to the grid by clamp bars.

(8) Vibrating screens are usually located at a high level on the plant and in dry climates will produce a continuous plume of wind-borne dust. As with cone crushers there is often a "chimney" effect through discharge chutes and a resulting up-current of air through the decks. The vibrating screens will often blanket the entire down-wind working area with dust and on a mobile plant the only practicable alleviation is a water-spray bar at the feed end. The finest possible spray should be generated and the nozzles should be so positioned that the incoming aggregate falls through the largest practicable volume of "mist". The objective here is to coat as many aggregate surfaces as possible before the adhering dust can be displaced into the atmosphere.

(9) Stock-piles present their own dust problems:

   (a) If the incoming stone is dry, dust particles will become wind-borne from virtually any size of aggregate leaving the conveyor head although aggregate sizes 0 − 6 mm present the main problem. If the conveyor discharge height is adjustable, free-fall can be minimised; otherwise water sprays as described above can be used.

   (b) Stock-piles of aggregates in the range of 0 − 6 mm can be water-sprayed. This will form a wet skin on the surface and prevent dust being taken by the wind. Of course, the skin will be broken by the loading shovel and dust will be freed. More water can be used but may make unacceptable the cost of subsequently drying the stone, for example in a coating process.

Whatever steps are taken by way of dust suppression or containment there should always be a good supply of dust respirators (masks) and a continuing supply of renewable filters. The respirators should comply with British Standard Specification BS 2091:1969 or its equivalent and the use of pre-filters is recommended where conditions are particularly arduous.

PORTABLE CONVEYORS

This section relates specifically to machines produced by Universal Conveyor Co Ltd, and is extracted with acknowledgement from that company's service literature.

Chain drives (transmission)
Chains are fitted fairly tightly at installation with only a small amount of slack.

New chains will loosen up slightly but this initial elongation is far greater than that which occurs during the remainder of the chain life. After the first few weeks of operation, adjust the centre distance, especially on long drives. The chain must at all times be correctly tensioned. Efficient lubrication is essential as a well lubricated chain has a wear life far in excess of one not lubricated.

Lubrication oil/grease as recommended below or equivalent, should be applied to the chain, making sure that the oil is applied to the complete length of chain and that the oil is applied to the outside and inside plate edges, as oil applied on the centre line of the rollers cannot reach the pin-bush joints and is thus useless for retarding chain wear.

The chain and sprockets should be examined at least once a month.

Any dirt, etc. should be removed from the chain, sprockets and inside the chain guard.

Burrs or rust on the sprockets should be removed by wire brush, file, or both.

If the chain has become contaminated, or every 1,500 hours working, it should be removed and thoroughly cleaned by scrubbing in paraffin, allowed to dry and re-lubricated with the correct oil. If grease is used, it should be slightly heated and the chain fully immersed in it for five minutes. The chain should then be removed, allowed to dry and the surplus grease wiped away.

Recommended lubricants

Application	Lubricant
Normal temperature − slow speeds	Shell "Valvata" oil J.82
Normal temperature − normal speeds	Shell "Tonna" T.71
	Shell "Alvania" EP.1
	Shell "Alvania" R.1
High temperature (150°C) normal speeds	Shell "Tilvela" oil 75
Normal temperature, shockloading, normal speeds	Shell "Macoma" R.85

Electric motors
Electric motors are essentially reliable machines and require little maintenance, in fact they tend to suffer from overattention more than inattention. However, it is necessary to inspect the motor at regular intervals, the frequency depending upon the environment, but in the region of four to eight weeks would be a reasonable interval.

Avoid wherever possible:-
1) Damp and falling moisture (unless the motor is specifically protected against this).
2) Dirt, especially fluff which may cause blocked ventilation.
3) Obstructing or covering motor.
4) Excessive heat. Motors should not be located in an area where the ambient temperature exceeds 40°C, unless specifically designed to do so.

Routine work
1) Remove the fan cover and ensure that all air inlet holes are completely free. Clean away any dirt, fluff etc. from behind the fan and along the ribs of the frame.
2) Commutator motors should have their brushes and commutator checked for signs of wear. When renewing brushes, care should be taken to ensure that the correct grade and size of brush is used.

Lubrication

Motors equipped with grease fittings and relief plugs should be re-lubricated by the following procedure.

1) Wipe clean the pressure-gun fitting and the regions around the motor grease fittings.
2) Remove the relief plug and free the relief hole of any hardened grease.
3) Add grease with the motor at a standstill until new grease is expelled through the relief hole. Greasing the motor at a standstill will minimise the possibility of grease leakage along the shaft seals.
4) Run the motor for about ten minutes with the relief plug removed to expel excess grease.
5) Clean and replace the relief plug.

The entry of a little grit can ruin a bearing.

Motors with sealed bearings should not require greasing.

Bearings

Self-lube bearings are filled with the correct amount of grease, sealed for life, and do not usually require relubricating unless operating under extremes of speed and loading, wet or dirty conditions, or extended running.

Most self-lube bearings have relubrication facilities and when necessary can be made re-greasable by removing the plastic plug and fitting a grease nipple.

Re-greasable bearings are supplied charged with grease. Re-grease by removing top of bearing housing or through grease nipple if fitted.

The recommended lubricant for re-greasable bearings is Shell Alvania R2 or R3 for temperature range −35°C to 100°C/135°C.

Do not over grease the bearings as this causes rapid temperature rise resulting in wear.

All bearings should be checked regularly for wear and damage and replaced as necessary. Keep all bolts securely tightened.

Grease gun fittings and grease nipples must be wiped clean immediately before re-greasing as the entry of a little grit can ruin a bearing.

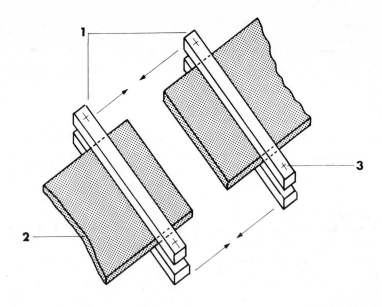

1. 4" × 2" TIMBERS OR SIMILAR 3. CLAMPING BOLT EACH END
2. BELT

Fig. 40 SIMPLE CONVEYOR BELT STRETCHERS

Belt joining

Metal fasteners
1. With the aid of belt stretchers, draw both ends of belt together until belt is reasonably tight. Mark uncut end where squared end overlaps and cut off surplus belt, again ensuring that cut is square to the centre line. Cut back corners as shown in Fig. 41. The belt fasteners may then be fitted as per manufacturer's instructions. Remove belt stretchers.
2. Tension the belt in accordance with the instructions on the tension gear sheet.
3. Before starting the conveyor, refer to the "Belt Tracking" paragraph.

H.D. type conveyor belt fasteners
H.D. No. 1 for use with belts $\frac{1}{4}"$ to $\frac{7}{16}"$ thick.
H.D. No. 1$\frac{1}{2}$ for use with belts $\frac{3}{8}"$ to $\frac{11}{16}"$ thick.
H.D. No. 2 for use with belts $\frac{1}{2}"$ to $\frac{11}{16}"$ thick.

98

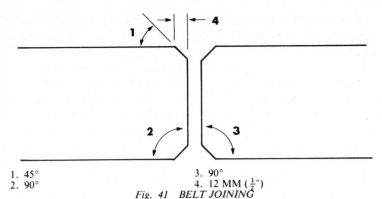

1. 45°	3. 90°
2. 90°	4. 12 MM ($\frac{1}{2}$")

Fig. 41 BELT JOINING

Fitting instructions

Ensure that the belt ends are square then supporting the belt with a wooden plank nail template in position with the belt ends held tightly against the lugs. Punch the holes with the belt punch. Remove template and insert bottom plate assemblies from underside of belt. Fit top plates and nuts, run down evenly and tighten. Break off unwanted bolt ends with bolt breaker. Retighten nuts after a few hours running.

Belt tracking

If the belt tends to run to one side it is essential the fault be corrected immediately to prevent the possibility of damage to the belt, particularly the edges. The conveyor should be inspected to ensure all the drums and idlers, on both the carrying and return strands of the belt, are in position and freely turning. Check side roller "lead" is in correct direction as shown in the sketch below. Any damaged or worn drums or rollers should be replaced. Check drums for material build-up and clean if necessary. Should the belt still run to one side, retracking should be carried out as follows:

BY MOVING THE LEFT SIDE OF THE BASE PLATE FORWARD, THE BELT WILL BE TRACKED TO THE RIGHT

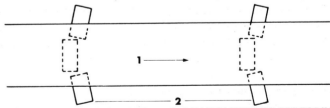

1. DIRECTION OF BELT TRAVEL   2. LEAD ON SIDE IDLERS FORWARD
BY MOVING THE RIGHT SIDE OF THE BASE PLATE, THE BELT WILL BE TRACKED TO THE LEFT
Fig. 42

99

Ensure side of belt and conveyor structure is kept clean and material is not allowed to build-up on the structure, drums, idlers, belt, scraper or chutes etc. Keep the feed centralised on the belt. To ignore these instructions could result in belt damage.

Belt scrapers
Inspect regularly and ensure the scraper blade only is in contact with the belt. Never let the metal frame scrape the belt as severe damage to the belt will result, particularly at the fasteners.

Tension gear

Screw take-up
The belt is tensioned by operating the adjusting screws. It is essential the drum remains square to the frame when tensioning is completed. This is achieved by ensuring the drum is first made square to the frame and then an equal amount of movement applied to each bearing in turn, until the correct belt tension is achieved. It is advisable to move each bearing a small distance at a time to prevent excess strain being applied to the belt and drum shaft bearings.

Correct tension is achieved when the drive drum starts the belt and continues it moving when loaded without any slip occuring. The tension screws should be kept clean and well greased.

When all the screw tension has been taken up it is time for the belt to be shortened.

Anti-runback gear

Anti-runback gear – installing
1. Before starting a conveyor motor for the initial test run, it is essential for the anti-runback gear either to be made inoperable, removed from the gear unit, or the vee drive, or chain drive disconnected from the motor pulley. This will allow a check to be made to ensure the motor shaft is rotating in the correct direction when started.
2. The anti-runback gear can then be checked to ensure it is fitted correctly.

Anti-runback fitted into gearboxes
1. These normally require no maintenance as they are lubricated within the gearbox.
2. A check should be carried out periodically to ensure the unit is still functioning correctly.
3. If the unit is used regularly, an inspection of the locking rollers should be made. Extreme care should be taken when removing a unit from the gearbox to ensure no rollers or springs are lost or damaged. Worn parts should be replaced as required. When re-fitting, it is recommended the maker's instructions be strictly adhered to.

Anti-runback fitted to head shaft
1. Regular inspection of the internal parts is recommended.
2. Fit and lubricate in accordance with the instructions given on the plate fitted to the cover.

Wheels
1. Fabricated steel.
   Inspect regularly and either repair any damage as quickly as possible or replace wheel as required.
   Refer to the above for bearing maintenance.
2. Pneumatic tyred.
   Regularly check and maintain the recommended tyre pressure.
   Remove stones or any material which has become embedded in the tyre or tread.
   Should the side wall be damaged, it is advisable to seek the advice of a tyre specialist to determine whether or not the tyre should be replaced immediately.
   Refer to the above for bearing maintenance.

Castors
1. Inspect for wear or damage and either repair or replace as required.
2. Lubricate wheels and hinge pins regularly.
3. Ensure fixing bolts and wheel bolts are maintained tight.

Suggested towing restrictions
1. Under no circumstances should a portable conveyor be towed on a site at a speed in excess of 4 miles per hour.

2. Care should be taken to ensure the ground is level in order to avoid the possibility of the conveyor tipping over during the towing operation or after removal of the towing vehicle.
3. Remove any obstructions such as boulders, etc., from the path of the conveyor.
4. Make sure the operators are sufficiently competent to carry out the towing operation without mishap to the conveyor or personnel.
5. In the case of an adjustable boom it is essential the boom is adjusted to its lowest position before towing commences.
6. Care should be taken to ensure the ground in the towing area is firm and suitable for carrying the load on each wheel.
7. Ensure the towing vehicle is suitable for the duty.
8. The operator must make himself aware of and observe all Ministry of Transport or any other relevant body regulations appertaining to towing vehicles.

Feed boot
Should be regularly checked to ensure it is feeding material centrally on to belt. Suitable adjustments should be made if necessary to achieve a satisfactory condition. Worn parts should be rectified as soon as possible and any obstruction preventing the free movement of material removed.

Sealing rubbers should be adjusted as frequently as required and replaced as necessary.

Chutes
Inspect regularly and repair or replace any damaged or worn parts as required. Remove any obstruction preventing the material being smoothly discharged.

Guards
Should be kept in position and well maintained. Any damage should be repaired immediately.

Troughing and parallel rollers
These are fitted with grease packed sealed bearings and require no greasing or oiling. They should be inspected regularly to ensure they are free to rotate, any obstruction should be

removed and the rollers in the vicinity examined for excessive wear.

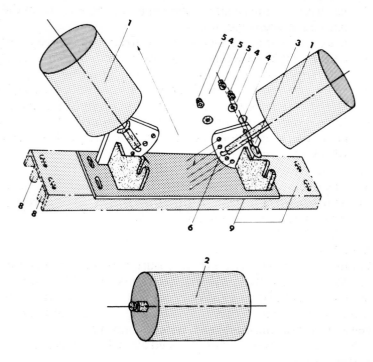

1. SIDE ROLLER AND SPINDLE   2. CENTRE ROLLER AND SPINDLE
3. GALVANIZED 'U' BOLT 4. PLAIN WASHER 5. LOCKNUT 6. 'L' BOLT
7. BELT TRAVEL   8. DISTANCE PIECES REQUIRED ON TRANSOME
TYPE BASE ONLY   9. TRANSOME TYPE BASE SHOWN IN CHAIN DOT,
BASE PLATE TYPE SHOWN IN FULL LINE
Fig 43   ASSEMBLY OF MULTITROUGH IDLERS

Suggested site stocks spare parts for portable conveyors
Bearings for snub shafts.
Bearings for head shafts.
Bearings for tail shafts.
Drive chains.
Belt fasteners.
Impact idlers.
Troughing idlers.
Parallel idlers.

103

Suggested preventive maintenance schedule for conveyors

Daily:
(a) Check that all troughing and return idler rollers are rotating freely. Replace any seized rollers.
(b) Check conveyor belt joints.
(c) Check tracking of conveyor belt and retrack if necessary.
(d) Check that belt scraper is effective.
(e) Check that all guards are in position and secure.

Weekly:
(a) Grease bearings of tail shaft, head shaft and snub shaft. (Where not self-lub.)
(b) Check tension of conveyor belt. Adjust if necessary.

Monthly:
(a) Check vee belt tension. Adjust if necessary. (Not portable conveyors)
(b) Check gearbox oil level. Top up if necessary.
(c) Check, adjust if necessary replace the rubber blade of the belt scraper.
(d) Replace any worn or ineffective skirt rubbers.
(e) Check and if necessary adjust drive chain tension. (Portable conveyors).
(f) Check that anti-runback device (if fitted) is effective.

6 monthly:
(a) Drain, flush and refill oil lubricated gearboxes.

APPENDIX: MISCELLANEOUS TABLES AND INFORMATION

Boiling point of water at atmospheric pressure = 100C° (212°F)
Freezing point of water = 0C° (32°F)
°C = $\frac{5}{9}$ (°F − 32)
°F = $\frac{9}{5}$ °C + 32
1 cubic foot of water at 4°C weighs 62.428 lbs

1 hp = 550 ft. lbs/sec. = 33000 ft. lbs/min. = 745.5 watt

π = 3.1416
Circumference of circle = π × diameter = πd
Area of circle = π × radius squared = πr^2

Volume of cone = area of base (πr^2) × $\frac{1}{3}$ perpendicular height
(Useful for stockpile calculations)

Area of triangle = base × $\frac{1}{2}$ perpendicular height
Area of rectangle = length × width

Conversion tables

Length
Inches × 25.4 = Millimetres
Inches × 2.54 = Centimetres
Feet × 30.48 = Centimetres
Feet × .3048 = Metres
Yards × .9144 = Metres
Miles × 1.6093 = Kilometres
Millimetres × .03937 = Inches
Centimetres × .3937 = Inches
Metres × 39.37 = Inches
Metres × 3.281 = Feet
Metres × 1.094 = Yards
Kilometres × 3280.9 = Feet
Kilometres × 1093.6 = Yards
Kilometres × .621 = Miles
1 millimicron = .001 micron
1 micron = .001 millimetre
Volume
Cu. in × 16.383 = Cu. centimetres
Cu. ft. × .0283 = Cu. metres
Cu. yds. × .7645 = Cu. metres
Cu. centimetres × .06102 = Cu. ins.
Cu. metres × 35.3145 = Cu. ft.
Cu. metres × 1.3079 = Cu. yards

Area

Sq. inches × 645.2 = Sq. millimetres
Sq. inches × 6.452 = Sq. centimetres
Sq. feet × .0929 = Sq. metres
Sq. yards × .8361 = Sq. metres
Acres × .4047 = Hectares
Acres × .00405 = Sq. kilometres
Sq. mile × 2.59 = Sq. kilometres
Sq. millimetres × .00155 = Sq. inches
Sq. centimetres × .155 = Sq. inches
Sq. metres × 10.764 = Sq. feet
Sq. metres × 1.196 = Sq. yards
Hectares × 2.471 = Acres
Sq. kilometres × 247.11 = Acres
Sq. kilometres × .3861 = Sq. miles

SCRATCH TESTS BY THE MOHS SCALE OF HARDNESS

Scale	Mineral	Scratch Test
1	Talc	(Softest)
2	Gypsum	
$2\frac{1}{2}$		Fingernail
3	Calcite	
4	Fluorite	
5	Apatite	
$5\frac{1}{2} - 6$		Knife blade or plate glass
6	Feldspar	
$6\frac{1}{2} - 7$		Steel file
7	Quartz	
8	Topaz	
9	Carborundum	
10	Diamond	

PHYSICAL PROPERTIES OF COMMON ROCKS

TYPE	GENERAL HARDNESS	AVERAGE WEIGHT		AVERAGE SPECIFIC GRAVITY	AVERAGE COMPRESSIVE STRENGTH		ABSORPTION %	ABRASION TESTS	
		lbs/cub. ft	kg./cub. m.		lbs/sq. inch	kg./sq. cm.		LOS ANGELES	TOUGHNESS
LIMESTONE	MEDIUM	95 (CRUSHED)	1522	2·63	17500	1230	0·61	33·8	8
DOLOMITE	MEDIUM	95 (CRUSHED)	1522	2·71	21200	1490	1·09	27·1	8
GRANITE	HARD	98 (CRUSHED)	1570	2·63	25000	1758	0·30	41·5	9
QUARTZITE	HARD	110 (CRUSHED)	1763	2·70	27000	1898	0·24 to 0·38	26·1 to 30·3	13 to 19
GRANITIC GRAVEL	VERY HARD	100	1603	2·65	25000 to 35000	1758 to 2460	0·30	VERY VARIABLE	

WEIGHTS OF STEEL SECTIONS
METRIC ROUNDS

Size (dia/mm)	Kg Per Foot	Kg Per Metre	Feet Per Tonne	Metres Per Tonne
2·0	·00750	·0246	133,300	40,600
2·4	·0108	·0354	92,600	28,200
2·5	·0117	·0385	85,300	26,000
2·8	·0147	·0482	68,000	20,700
3·0	·0169	·0554	59,300	18,000
3·2	·0192	·0630	52,100	15,900
3·5	·0230	·0754	43,500	13,270
4·0	·0300	·0984	33,300	10,160
4·2	·0331	·1085	30,200	9,210
4·4	·0363	·1191	27,500	8,400
4·5	·0380	·1246	26,300	8,030
4·8	·0432	·141	23,100	7,050
5·0	·0469	·153	21,300	6,500
5·2	·0507	·166	19,700	6,010
5·5	·0567	·186	17,600	5,370
5·8	·0631	·207	15,900	4,830
6·0	·0675	·222	14,800	4,520
6·2	·0721	·237	13,870	4,230
6·5	·0792	·260	12,620	3,850
6·8	·0867	·285	11,530	3,520
7·0	·0919	·301	10,880	3,320
7·5	·1055	·346	9,480	2,890
8·0	·1200	·394	8,330	2,540
8·5	·1355	·445	7,380	2,250
9·0	·152	·498	6,580	2,010
9·5	·169	·555	5,910	1,800
10·0	·188	·615	5,330	1,630
10·5	·207	·678	4,840	1,470
11·0	·227	·744	4,410	1,343
11·5	·248	·814	4,030	1,229
12·0	·270	·886	3,700	1,129
12·5	·293	·961	3,410	1,040
13·0	·317	1·040	3,160	962
13·5	·342	1·121	2,930	892
14·0	·368	1·206	2,720	829
14·5	·394	1·293	2,540	773
15·0	·422	1·384	2,370	722
15·5	·451	1·48	2,220	677
16·0	·480	1·57	2,080	635
16·5	·511	1·67	1,960	597
17·0	·542	1·78	1,850	562
17·5	·574	1·88	1,740	531
18·0	·608	1·99	1,650	502
18·5	·642	2·11	1,560	475
19·0	·677	2·22	1,480	450
19·5	·713	2·34	1,400	427
20·0	·750	2·46	1,333	406
20·5	·788	2·59	1,269	387
21·0	·827	2·71	1,209	369
21·5	·867	2·84	1,154	352
22·0	·908	2·98	1,102	336
22·5	·949	3·11	1,053	321
23·0	·992	3·25	1,008	307
23·5	1·036	3·40	966	294
24·0	1·080	3·54	926	282

METRIC ROUNDS

Size (dia/mm)	Kg Per Foot	Kg Per Metre	Feet Per Tonne	Metres Per Tonne
24·5	1·126	3·69	888	271
25·0	1·172	3·85	853	260
26	1·268	4·16	789	240
27	1·367	4·48	732	223
28	1·47	4·82	680	207
29	1·58	5·17	634	193
30	1·69	5·54	593	181
31	1·80	5·91	555	169
32	1·92	6·30	521	159
33	2·04	6·70	490	149
34	2·17	7·11	461	141
35	2·30	7·54	435	132·7
36	2·43	7·97	411	125·4
37	2·57	8·42	390	118·7
38	2·71	8·88	369	112·6
39	2·85	9·36	351	106·9
40	3·00	9·84	333	101·6
41	3·15	10·34	317	96·7
42	3·31	10·85	302	92·1
43	3·47	11·38	288	87·9
44	3·63	11·91	275	84·0
45	3·80	12·46	263	80·3
46	3·97	13·02	252	76·8
47	4·14	13·59	241	73·6
48	4·32	14·2	231	70·6
49	4·50	14·8	222	67·7
50	4·69	15·4	213	65·0
52	5·07	16·6	197	60·1
54	5·47	17·9	183	55·7
55	5·67	18·6	176	53·7
56	5·88	19·3	170	51·8
58	6·31	20·7	159	48·3
60	6·75	22·1	148	45·2
62	7·21	23·6	138·7	42·3
64	7·68	25·2	130·2	39·7
65	7·92	26·0	126·2	38·5
66	8·17	26·8	122·4	37·3
68	8·67	28·4	115·3	35·2
70	9·19	30·1	108·8	33·2
72	9·72	31·9	102·9	31·4
74	10·27	33·7	97·4	29·7
75	10·55	34·6	94·8	28·9
76	10·83	35·5	92·3	28·1
78	11·41	37·4	87·7	26·7
80	12·00	39·4	83·3	25·4
82	12·61	41·4	79·3	24·2
85	13·55	44·4	73·8	22·5
88	14·5	47·6	68·9	21·0
90	15·2	49·8	65·8	20·1
92	15·9	52·1	63·0	19·2
95	16·9	55·5	59·1	18·0
98	18·0	59·1	55·5	16·9
100	18·8	61·5	53·3	16·3
105	20·7	67·8	48·4	14·7
110	22·7	74·4	44·1	13·43
115	24·8	81·4	40·3	12·29

METRIC ROUNDS

Size (dia/mm)	Kg Per Foot	Kg Per Metre	Feet Per Tonne	Metres Per Tonne
120	27·0	88·6	37·0	11·29
125	29·3	96·1	34·1	10·40
130	31·7	104·0	31·6	9·62
135	34·2	112·1	29·3	8·92
140	36·8	120·6	27·2	8·29
145	39·4	129·3	25·4	7·73
150	42·2	138·4	23·7	7·22
155	45·0	148	22·2	6·77
160	48·0	157	20·8	6·35
165	51·0	167	19·6	5·97
170	54·2	178	18·6	5·62
175	57·4	188	17·4	5·31
180	60·8	199	16·5	5·02
185	64·2	211	15·6	4·75
190	67·7	222	14·8	4·50
195	71·3	234	14·0	4·27
200	75·0	246	13·33	4·06

METRIC SQUARES

Size A/F (mm)	Kg Per Foot	Kg Per Metre	Feet Per Tonne	Metres Per Tonne
4	·0382	·1253	26,200	7,980
4·5	·0484	·159	20,700	6,310
5	·0597	·196	16,800	5,110
5·5	·0722	·237	13,850	4,220
6	·0860	·282	11,630	3,550
6·5	·1009	·331	9,910	3,020
7	·1170	·384	8,550	2,610
7·5	·1343	·441	7,450	2,270
8	·153	·501	6,540	2,000
8·5	·173	·566	5,800	1,770
9	·193	·635	5,170	1,580
9·5	·215	·707	4,640	1,410
10	·239	·783	4,190	1,277
10·5	·263	·864	3,800	1,158
11	·289	·948	3,460	1,055
11·5	·316	1·036	3,170	966
12	·344	1·128	2,910	887
12·5	·373	1·224	2,680	817
13	·404	1·324	2,480	756
13·5	·435	1·43	2,300	701
14	·468	1·54	2,140	652
14·5	·502	1·65	1,990	607
15	·537	1·76	1,860	568
16	·611	2·01	1,640	499
17	·690	2·26	1,450	442
18	·774	2·54	1,293	394
19	·862	2·83	1,160	354
20	·955	3·13	1,047	319
21	1·053	3·45	950	289
22	1·156	3·79	865	264
23	1·263	4·14	792	241
24	1·376	4·51	727	222
25	1·49	4·90	670	204
26	1·61	5·30	620	189
27	1·74	5·71	574	175
28	1·87	6·14	534	163
29	2·01	6·59	498	152
30	2·15	7·05	465	142
31	2·29	7·53	436	132·9
32	2·45	8·02	409	124·7
33	2·60	8·53	385	117·3
34	2·76	9·05	362	110·5
35	2·93	9·60	342	104·2
36	3·09	10·15	323	98·5
37	3·27	10·72	306	93·3
38	3·44	11·31	290	88·4
39	3·63	11·91	275	83·9
40	3·82	12·53	262	79·8
41	4·01	13·17	249	75·9
42	4·21	13·82	237	72·4
43	4·42	14·5	227	69·1
44	4·62	15·2	216	65·9
45	4·84	15·9	207	63·1
46	5·05	16·6	198	60·4
47	5·28	17·3	190	57·8
48	5·50	18·0	182	55·4

METRIC SQUARES

Size A/F (mm)	Kg Per Foot	Kg Per Metre	Feet Per Tonne	Metres Per Tonne
49	5·73	18·8	174	53.2
50	5·97	19·6	168	51.1
55	7·22	23·7	138·5	42.2
60	8·60	28·2	116·3	35·3
65	10·09	33·1	99·1	30·2
70	11·70	38·4	85·5	26·1
75	13·43	44·1	74·5	22·7
80	15·3	50·1	65·4	20·0
85	17·3	56·6	58·0	17·7
90	19·3	63·4	51·7	15·8

METRIC FLATS—Kg PER METRE

Thickness T (mm) \ Width W (mm)	12	20	25	32	40	50	65	80	90	100	125	160
3	·282	·470	·587	·752	·940	1·175	1·53	1·88	2·11	2·35	2·94	3·76
5	·470	·783	·979	1·253	1·57	1·96	2·55	3·13	3·52	3·92	4·90	6·27
6	·564	·940	1·175	1·50	1·88	2·35	3·05	3·76	4·23	4·70	5·87	7·52
10	·940	1·57	1·96	2·51	3·13	3·92	5·09	6·27	7·05	7·83	9·79	12·53
12		1·88	2·35	3·01	3·76	4·70	6·11	7·52	8·46	9·40	11·75	15·0
16		2·51	3·13	4·01	5·01	6·27	8·15	10·03	11·28	12·53	15·7	20·1
20			3·92	5·01	6·27	7·83	10·18	12·53	14·10	15·7	19·6	25·1
25				6·27	7·83	9·79	12·73	15·7	17·6	19·6	24·5	31·3
30					9·40	11·75	15·3	18·8	21·1	23·5	29·4	37·6
40						15·7	20·4	25·1	28·2	31·3	39·2	50·1
50							25·5	31·3	35·2	39·2	49·0	62·7
65								40·7	45·8	50·9	63·6	81·4
75									52·9	58·7	73·4	94·0

METRIC ANGLES

Size	Kg per Foot	Kg per Metre	Ft per Tonne	Metres per Tonne
16 × 16 × 3	·208	·681	4810	1470
20 × 20 × 3	·265	·869	3770	1150
25 × 25 × 3	·337	1·104	2970	906
25 × 25 × 5	·537	1·76	1860	568
32 × 32 × 4	·573	1·88	1740	532
32 × 32 × 5	·704	2·31	1420	433
40 × 40 × 4	·726	2·38	1380	420
40 × 40 × 5	·896	2·94	1120	341
40 × 40 × 6	1·060	3·48	943	288
50 × 50 × 6	1·347	4·42	743	226

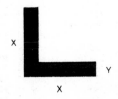

$$Area = 2XY - Y^2$$

Kg per Foot	Kg per Metre	Ft per Tonne	Metres per Tonne
·002388 × Area	·007833 × Area	418800 − Area	127700 − Area

114

EQUIVALENT TEMPERATURES
CENTIGRADE AND FAHRENHEIT

C°	F°	C°	F°	C°	F°	C°	F°
0	32	45	113	91	195·8	137	278·6.
1	33·8	46	114·8	92	197·6	138	280·4
2	35·6	47	116·6	93	199·4	139	282·2
3	37·4	48	118·4	94	201·2	140	284
4	39·2	49	120·2	95	203	141	285·8
5	41	50	122	96	204·8	142	287·6
6	42·8	51	123·8	97	206·6	143	289·4
7	44·6	52	125·6	98	208·4	144	291·2
8	46·4	53	127·4	99	210·2	145	293
9	48·2	54	129·2	100	212	146	294·8
10	50	55	131	101	213·8	147	296·6
11	51·8	56	132·8	102	215·6	148	298·4
12	53·6	57	134·6	103	217·4	149	300·2
13	55·4	58	136·4	104	219·2	150	302
14	57·2	59	138·2	105	221	151	303·8
15	59	60	140	106	222·8	152	305·6
15·5	60	61	141·8	107	224·6	153	307·4
16	60·8	62	143·6	108	226·4	154	309·2
17	62·6	63	145·4	109	228·2	155	311
18	64·4	64	147·2	110	230	156	312·8
19	66·2	65	149	111	231·8	157	314·6
20	68	66	150·8	112	233·6	158	316·4
21	69·8	67	152·6	113	235·4	159	318·2
22	71·6	68	154·4	114	237·2	160	320
23	73·4	69	156·2	115	239	161	321·8
24	75·2	70	158	116	240·8	162	323·6
25	77	71	159·8	117	242·6	163	325·4
26	78·8.	72	161·6	118	244·4	164	327·2
27	80·6	73	₁03·4	119	246·2	165	329
28	82·4	74	165·2	120	248	166	330·8
29	84·2	75	167	121	249·8	167	332·6
30	86	76	168·8	122	251·6	168	334·4
31	87·8	77	170·6	123	253·4	169	336·2
32	89·6	78	172·4	124	255·5	170	338
33	91·4	79	174·2	125	257	171	339·8
34	93·2	80	176	126	258·8	172	341·6
35	95	81	177·8	127	260·6	173	343·4
36	96·8	82	179·6	128	262·4	174	345·2
37	98·6	83	181·4	129	264·2	175	347
38	100·4	84	183·2	130	266	176	348·8
39	102·2	85	185	131	267·8	177	350·6
40	104	86	186·8	132	269·6	178	352·4
41	105·8	87	188·6	133	271·4	179	354·2
42	107·6	88	190·4	134	273·2	180	356·2
43	109·4	89	192·2	135	275	181	357·8
44	111·2	90	194	136	276·8	182	359·6

POUNDS PER SQUARE INCH CONVERSION

Lb. per sq. in.	Atmos-pheres	Feet of Water	Kg. per sq. c.m.	Lb. per sq. in.	Atmos-pheres	Feet of Water	Kg. per sq. c.m.
10	0·68	23·12	0·70	100	6·80	231·2	7·03
20	1·36	46·24	1·41	200	13·60	462·4	14·06
30	2·04	69·36	2·11	300	20·41	693·6	21·09
40	2·72	92·48	2·81	400	27·21	924·8	28·12
50	3·4	115·61	3·52	500	34·01	1156·1	35·15
60	4·08	138·73	4·22	600	40·81	1387·3	42·18
70	4·76	161·85	4·92	700	47·62	1618·5	49·21
80	5·44	184·96	5·62	800	54·42	1849·6	56·24
90	6·12	208·09	6·33	900	61·23	2080·9	63·27

116

INCH-MILLIMETRE EQUIVALENTS

Inch-Millimetre Equivalents of Decimal and Common Fractions From 1/64 to 1 in.

Inch	$\frac{1}{2}$'s	$\frac{1}{4}$'s	8ths	16ths	32nds	64ths	Millimetres
						1	0.397
					1	2	0.794
						3	1.191
				1	2	4	1.588
						5	1.984
					3	6	2.381
						7	2.778
			1	2	4	8	3.175
						9	3.572
					5	10	3.969
						11	4.366
				3	6	12	4.762
						13	5.159
					7	14	5.556
						15	5.953
		1	2	4	8	16	6.350
						17	6.747
					9	18	7.144
						19	7.541
				5	10	20	7.938
						21	8.334
					11	22	8.731
						23	9.128
			3	6	12	24	9.525
						25	9.922
					13	26	10.319
						27	10.716
				7	14	28	11.112
						29	11.509
					15	30	11.906
						31	12.303
	1	2	4	8	16	32	12.700
						33	13.097
					17	34	13.494
						35	13.891
				9	18	36	14.288
						37	14.684
					19	38	15.081
						39	15.478
			5	10	20	40	15.875
						41	16.272
					21	42	16.669
						43	17.066
				11	22	44	17.462
						45	17.859
					23	46	18.256
						47	18.653
		3	6	12	24	48	19.050
						49	19.447
					25	50	19.844
						51	20.241
				13	26	52	20.638
						53	21.034
					27	54	21.431
						55	21.828
			7	14	28	56	22.225
						57	22.622
					29	58	23.019
						59	23.416
				15	30	60	23.812
						61	24.209
					31	62	24.606
						63	25.003
1	2	4	8	16	32	64	25.400